IN SERVICE

to

MATHEMATICS

The Life and Work of MINA REES

AMY SHELL-GELLASCH

Docent Press

DOCENT PRESS
Boston, Massachusetts, USA
www.docentpress.com

Docent Press publishes books in the history of mathematics and computing about interesting people and intriguing ideas. The histories are told at many levels of detail and depth that can be explored at leisure by the general reader.

Cover design by Brenda Riddell, Graphic Details, Portsmouth, New Hampshire.

To $\mathcal{Z}$, the Wife of $\mathcal{Z}$, the children of $\mathcal{Z}$ and the son-in-law and ultimate grandson of $\mathcal{Z}$[1].

Acknowledgements

I would like to thank my doctorial advisor, Bill Howard, for his time, energy, and sense of humor, and for suggesting this topic. I thank my doctorial committee, Calixto Calderon, Fred Rickey, Martin Tangora, and Kristen Wottreng for their time and support. I would also like to thank the following people for their interest, help, and input and most importantly their time. Susan Newman, Chief Librarian of the Mina Rees Library. Alexandra Robinson and the staff in the Office of the President of The Graduate Center, CUNY. Jane House, Director of Publications, The Graduate Center, CUNY. Ray Ring, Director of building design and exhibitions, CUNY. John Rothman, GSUC Archivist. Joan Byers, Mina Rees's executive assistant at CUNY. Carol Mead and Ralph Elder (formerly) of the MAA Archives. Benjamin Stone of the University of Chicago Archives. Alison Oswald of the Smithsonian Institute Archives. Abbie Weinberg, NYU Archives. Also Jeanne LaDuke, Karen Parshall, Saunders Mac Lane, Dan Bernstein, and Burt Fein. And finally, my editor Scott Guthery, for seeing the value in this work and making sure it saw the light of day.

Preface

Mina Spiegel Rees (1902–1997) had a profound effect on mathematical research in the mid-twentieth century. Her tenure as the head of the Mathematical Sciences Division of the Office of Naval Research during its inception after World War II significantly influenced mathematical research in the United States. Her leadership in establishing the Graduate School and University Center of the City University of New York during the 1960s shaped graduate education at those institutions, as well as nationwide.

I arrived at the University of Illinois at Chicago in January 1998 to pursue a Doctor of Arts degree in mathematics. As I looked around for an advisor and thesis topic, Bill Howard gave me an obituary of Mina Rees he had cut out of the newspaper that fall. He said emphatically that someone had to do work on Mina Rees. In very short order Bill and I decided to work together and that Mina would be my topic. Surprisingly, my career path has had many parallels with Rees's.

This book explores in depth her thesis work in associative division algebras under L. E. Dickson at the University of Chicago in 1931; her work on the Applied Mathematics Panel during World War II; her work at the Office of Naval Research until 1953; and finally, her work in graduate education at the City University of New York.

Material for the book was gathered from original sources, articles about Rees and her work, records of interviews with Rees, interviews and correspondence with her colleagues, her personal papers, and related historical and mathematical material.

Contents

CHAPTER 1

Introduction

"Any respectable description of what Mina Rees has done, rather than an enumeration of the positions she has held, will inevitably read like a concise history of American mathematics during the last few decades."[1]

F. Joachim Weyl
Science, 1970

Rees was instrumental in facilitating scientific research during World War II. She revolutionized the form and scope of federal support of scientific research. She facilitated much of the development of early computers. And she ushered in a new era in graduate education at the City University of New York and influenced graduate education in general. All this she did in her official capacities as administrator and educator. However, she also had a profound effect on research and education through her personal involvement. She was dedicated to creating an environment in the United States in which the sciences, and mathematics in particular, grew and flourished. Moreover, she improved the working conditions of researchers and the learning environment of students. She was particularly interested and active in promoting women's education.

Her efforts both during the war and immediately following were recognized by the Mathematical Association of America (MAA) when it awarded its first Award for Distinguished Service to Mathematics to Rees in 1961. They stated, in part that, "Rees firmly built into the permanent structure and policies of the Office of Naval Research (ONR) the principle that the full scope of mathematics should form part of the total scientific effort properly supported by government sponsored research programs." The award citation goes on to note that Rees

---

[1][**91**, p. 1149].

had "contributed perhaps more than any other single person to the scope and
wealth of present day mathematical research activity in the United States."[2]

Research, both federal and academic, owes much of its present system to
the policies of Rees. She can be credited with helping to bring about the birth
of applied mathematics and the marriage of scholarly research with industrial,
social, and national needs. The organization of academic research and graduate
study of the second half of the twentieth century, particularly in the sciences,
might be quite different if not for her influence.

Simply reading the list of organizations[3] she contributed her time to and
helped govern gives a powerful sense of a life dedicated to the betterment of this
country. And the list of her awards[4] speaks to her competence and dedication
to science. She did all of this with an openness of character and warmth that
many enjoyed. Yet at the same time she was goal oriented and unwavering.

A description of Rees is better left to one of her biographers who was
fortunate enough to have known her personally. Uta Merzbach spoke of Rees at
the time of her death in an article in the *Notices of the American Mathematical
Society* (AMS).

> Mina Rees was eminently rational. Her devotion to reason
> helped her formulate goals clearly and allocate resources judi-
> ciously in accordance with these goals ... She could be a tough
> contract administrator. There were engineers who never for-
> gave the woman whom they regarded as the epitome of the
> autocratic Washington administrator with little understanding
> of cost overruns. At the same time there were mathematicians
> who never forgot to praise this colleague, whom they saw as
> the revitalization of research in numerical analysis.
>   Mina Rees was eminently intelligent. She comprehended
> quickly, communicated effectively, and thought creatively. Her

---

[2]See Appendix E for the complete award and citation of the MAA award.

[3]Appendix J for a partial list of committees, etc. Rees served on.

[4]Appendix H for a partial list of honors awarded to Rees.

ability to attach realizable pieces of basic research to mission-oriented applications of mathematics did much to develop a broadened base of support for mathematicians' work.

Mina Rees was eminently civilized. Her diplomatic skills were considerable; her conversational technique bespoke her broad knowledge base as well as her wide interest in mathematical and non-mathematical topics. Experience and reflection led her to a balanced outlook on teaching and research, the arts and the sciences, long-range and short-range planning, and the obligations of the professional and the private life.[5]

There are many questions to ask about such a full life. How did she choose to pursue mathematics as a woman in the 1920s? How did she find herself, again as a woman, in a prominent, and traditionally male, government position during the Second World War? What brought about her change of interest from pure to applied mathematics? What took her back into academia in the mid 1950s? Overall, how did she make the choices she did, that lead to a life that accomplished so much and effected such change? Whatever that quality may be, mathematics is all the richer for having it in Mina Rees.

---

[5][**29**, p. 872].

CHAPTER 2

Mina Rees, 1902–1997

In 1882 Alice Loise Stackhouse (1870–1946) emigrated to Pennsylvania from England with her family. Ten years later she married an insurance salesman named Moses Rees (1858–1932). Moses Rees was a first generation German-American, born in New York.[1] They had five children, Elsa, Albert, Clyde, Clarence, and the youngest, Mina Spiegel Rees, was born on August 2, 1902, in Cleveland, Ohio. When Mina was two years old, Moses returned to New York with his family.

Rees was a naturally inquisitive and hard working child, receiving all A's in the public school she attended in the Bronx. In eighth grade one of her teachers, whom she described as her "current hero" at the time[2], suggested that she take the entrance exam for Hunter High School. Hunter High School, affiliated with Hunter College in the New York City public education system, was a tuition-free school for gifted young women. Rees was accepted, and graduated from Hunter High School valedictorian in 1919. While at the high school she followed a college preparatory sequence including four years of mathematics which she very much enjoyed. Rees continued on to Hunter College, where she chose to pursue mathematics, "[not] because of its practical uses at all; it was because it was such fun."[3] She was also interested in becoming a lawyer and so took several history courses. But when it came time to register every semester, she could not resist the mathematics courses. Her family did not pressure her to follow a more traditional course of study. She explained that in her family, they

---

[1][**30**].

[2][**9**, p. 258].

[3]*Ibid.*, p. 258

did not interfere in each other's lives.[4] Rees was the only person in her family to pursue a mathematical career.

Since Hunter High School and Hunter College were female institutions with primarily female faculty, Rees was not aware of any gender gap nor discrimination in her classes. She had never considered the idea that mathematics might be considered (at the time) harder for women.

> Now, by definition, a college has a mathematics department, so at a college for women, you have women in mathematics! ... So I didn't meet any discouragement at all when I was going into math. Indeed, I didn't know it was a peculiar thing to do. I did what everybody did: pick that field that I found most interesting and decided to major in it. It never occurred to me that there was anything the matter with that.[5]

Along with her growing love of mathematics, Rees developed very strong organizational and social skills. She was student body president and editor of the yearbook, as well as a member of Phi Beta Kappa and Pi Mu Epsilon honor societies. She also received the prestigious H Pin for her contribution to the college as the President of the Student Self-Governing Association (student council). She graduated summa cum laude in 1923. At the centennial celebration of Hunter College in 1970, Rees was named one of the twelve most distinguished graduates. And in 1973, she was inducted into the Hunter College Hall of Fame by the alumni association.[6]

During her freshman year at Hunter College, the college asked her to teach a trigonometry laboratory course in surveying, called 'transit'. In order to do the best job possible, she took summer courses at Columbia University in education. Mina then taught at Hunter College for the next three years, while still an undergraduate there, though at half the beginning teacher's salary.

Upon graduation, Hunter College asked Rees to join the faculty. However, true to Rees's desire for excellence, she did not feel she had the mathematical

---

[4]*Ibid.,* p. 259.

[5][**9**, p. 258].

[6]Rees later served as president of the alumni association.

background to be a good college instructor. Instead she accepted a teaching position at Hunter High School and returned to Columbia as a full-time graduate student in mathematics. In a 1969 interview with Uta Merzbach, Rees described her experiences at Columbia:

> When I was graduated from Hunter College, I was offered a job at the College, but I had formed a firm opinion when I was an undergraduate that this was a bad mistake that the College was making, employing people who had just graduated. I felt that the standards of the College were not high enough and that people should be better educated before they ... became teachers there. So, I said I could not under any circumstances, teach at the College because I wasn't [well enough educated].
>
> The head of the department was appalled at anything like this, so she got me a job at Hunter High School, where I taught for three years while I got a master's degree at Columbia ... I really could be a full-time student at Columbia at the same time that I was teaching, because this was the easiest high school to teach in, ... entirely different from ... teaching in high school today. The students were a selected group of girls and very easy to teach.[7]

While an undergraduate, Rees had taken an abstract algebra course that she enjoyed, and hoped to pursue a Ph.D. in algebra at Columbia. Unfortunately, after completing several of the required math courses, she realized that she could not pursue her doctorate at Columbia.

> When I had taken four of their six-credit graduate courses in mathematics and was beginning to think about a thesis, the word was conveyed to me – no official ever told me this, but I learned – that the Columbia mathematics department was really not interested in having women candidates for Ph.D.s.[8]

Rees goes on to say, "That was the only episode that raised a question about the appropriateness of mathematics as a field for women before I had my Ph.D.

---

[7][**39**].
[8][**9**, p. 258].

It was really traumatic for me." Rees adjusted her course of study toward a master's degree in education[9] and graduated in 1925 from Columbia Teachers College. She also continued to take a smattering of law courses.

Rees now took the lecturer position she had been originally offered at Hunter College. Though the position was a teaching position, Hunter encouraged professional growth through seminars and society meetings. By 1929 Rees had saved enough money to pursue her original goal of a Ph.D. in algebra. She took a leave of absence from Hunter and headed for The University of Chicago to study under Leonard Dickson. Chicago had a very strong mathematics department, headed by Gilbert A. Bliss. Chicago was also more accepting of women than other U.S. institutions, and over the next few decades, would produce a large number of the United States' female scientists.[10] Historian of mathematics Della Fenster suggests that, "...perhaps as a result of his [Dickson's] early reputation as an advisor for women, his institution [Chicago], and his strong research program, Dickson evolved into a notably successful advisor for women pursuing mathematics doctorates in the United States."[11] Between 1900 and 1939, Dickson advised 8% of all female Ph.D.s in mathematics in the United States, and 40% of those at the University of Chicago.[12]

Rees seems to have made a memorable impression on her fellow graduate students at Chicago. Saunders Mac Lane remembers her as very capable, and feels that her not continuing in research was a "lost opportunity" for mathematics.[13] Julia Bower, another fellow student shared this memory of Rees with to Saunders Mac Lane.

> Vibrant would be a word to describe her. She was happily and totally interested in the people around her and in all the events taking place. I sat next to her in one of Professor Lane's classes. While most of us were busily taking notes, she sat there concentrating every fiber of her mind on what was being said, only occasionally writing something or making an intelligent

---

[9]See Appendix A for Rees's course work at Columbia.
[10][**85**].
[11][**20**, p. 19].
[12]*Ibid.*, p. 14.
[13][**29**, p. 870].

remark or asking an intelligent question. I had the feeling that she was thoroughly absorbing the important points, that she would not need to do any further studying at all.

I believe that all her life she went directly to the heart of a problem and dealt with it efficiently. Yet she had patience with those who were not so bright. She really liked people and enjoyed being with them.[14]

Rees completed her Ph.D. in December 1931, with a thesis in associative division algebras, under the direction of Dickson. She then returned to New York to become an instructor at Hunter College. In 1932 she was promoted to Assistant Professor, and in 1940 to Associate Professor. She taught the whole range of undergraduate courses of that time including geometry, calculus, and what she termed experimental courses.

In the early 1930s, many mathematicians found it difficult to find work. For example, Saunders Mac Lane could only find a position teaching at the high school level. It was very lucky that when Rees headed to Chicago to obtain her Ph.D., she took a leave of absence from Hunter College instead of giving up her position entirely. For when she completed her degree the depression was in full swing and her job at Hunter may not still have been available. She devoted most of her time to teaching. Rose Donaldson, a former student of Rees's who graduated from Hunter College in 1936, remembers her well. "I definitely remember Professor Rees ... The courses were definitely theoretical ... She was an exciting teacher, enthusiastic, and involved with her work."[15]

Rees did not publish any mathematics during the ten years (1932–1942) she spent at Hunter. However, the editor of the book reviews for *Scripta Mathematica*, Lao G. Simons, was also on the Hunter faculty, so she wrote three reviews for him,[16] as well as several for the AMS and the MAA. There are several possible reasons why Rees did not continue her mathematical research.

---

[14]*Ibid.*, p. 869.

[15][**15**].

[16]Rees published three book reviews for *Scripta Mathematica* during this time. In 1935 she reviewed *Triumph Der Mathematik,* by Heinrich Dörrie, in 1936 she reviewed *The Search Far Truth,* by E. T. Bell, and in 1940 she reviewed *A Semicentennial History of the American Mathematical Society, 1888–1998,* by Raymond C. Archibald.

During this time, teaching loads at many American institutions were high, usually four courses per term. The push to publish was also not yet the norm. Finally, in an autobiographical piece in the *Newsletter of the Association for Women in Mathematics* in 1979, Rees wrote, "When I was young my only ambition was to be a research mathematician. In retrospect, I don't believe that was ever possible for me."[17] However, as a college student, when she thought of mathematics as a career, she thought of teaching. While back at Hunter as a professor, Rees also became increasingly involved with the Mathematical Association of America and the American Mathematical Society. Time pressures and her continuing desire to be involved, along with a possible realization that her talents lay more in teaching and administration, may all have contributed to her choice to not continue in mathematical research. "My course was set. I was committed to administration, not research, but administration with a heavy orientation toward science.[18]

In 1930, Rees attended her first mathematics conference at Brown University. At this meeting she met and dinned with Marston Morse and G. D. Birkhoff. These meetings caused her to become "enchanted with mathematicians."[19] She then attended as many meetings as possible in the New York area, meeting as many mathematicians as she could. "Though I was not a research mathematician and though I soon learned that I couldn't understand much at the meetings, I did find them useful in giving me some idea of the directions that mathematics research was taking."[20] Of particular import, Rees meet Richard Courant at a meeting while he was a visiting lecturer of the American Mathematical Society in New York. When Courant came to New York University in 1934, Rees continued their acquaintance, which would pay dividends in the future. The time spent at meetings was invaluable in directing and shaping the future of her career.

While at a party in 1936, Rees met Leopold Brahdy. He was a physician planning to travel to Russia with other researchers to study the effect of the policies of the Russian government on life in that country. Rees was planning on attending the Mathematical Congress in Oslo at the same time, and decided

---

[17][**77**, p. 18].
[18][**77**].
[19]*Ibid.*, p. 16
[20]*Ibid.*, p. 16

to travel with him. Born in 1892 in Vienna, Brahdy immigrated to the United States with his family at the age of six. Brahdy and Rees were married in 1955.[21] They enjoyed traveling together up until Brahdy's death in 1977.

In 1943, the Applied Mathematics Panel (AMP) was created by the U.S. Office of Science Research and Development (OSRD) to support war related research. Courant suggested Rees to AMP director Warren Weaver as someone who could not only help mathematically, but more importantly, administratively. In what would result in a significant career move, Rees once more took a leave of absence from Hunter College and became a Technical Aide and executive assistant to Warren Weaver at AMP. In these roles, she acted as the liaison between the research groups contracted by the AMP to do war related research, and the U.S. government.

With the close of World War II, scientific and mathematical research changed focus. Thus the Applied Mathematics Panel was disbanded in 1946 and its role was taken over by the new Office of Naval Research. Rees was immediately appointed to head its Mathematics Branch and moved to Washington D.C. In 1948, Rees and Weaver were both awarded the King's Medal of Britain and the Presidential Certificate from President Truman, for their efforts during the war.[22] By 1949 the Mathematics Branch had expanded into the Mathematical Sciences Division, with Rees still at its helm. By 1952 she was appointed Deputy Science Director of the ONR. Through her position at the ONR, Rees made policy and funding decisions that greatly influenced the nature of scientific and mathematical research in the United States. One researcher with whom Rees worked christened her "the angel of mathematics" for her work at ONR.[23]

In 1953, Rees left the ONR and Washington to return to Hunter College as Professor of Mathematics and Dean of Faculty. The city colleges of New York, namely City College, Hunter College, Brooklyn College, and Queens College, along with three community colleges, were brought together to form The City University of New York (CUNY) in 1961. At that time, Rees moved over to CUNY as Professor and Dean of Graduate Studies. She was the first female

---

[21]Though Rees meet Brahdy in 1936, they were not married until 1955, when Rees was in her early fifties. The couple had no children.

[22]See Appendices E and F for the full awards.

[23][30].

dean of a co-educational institution in the United States.[24] While Dean, she had two main goals, to establish solid Ph.D. programs in all fields of study, and to rework graduate education to fulfill the needs of a more diverse student population and employer base. Of particular interest to Rees was advancing opportunities for and acceptance of women in academia. In 1968, Rees became a Professor and Provost of the new Graduate Division of CUNY. The following year she became the first President of The Graduate School and University Center (GSUC) of CUNY. She quickly built GSUC into a respected graduate school.

During this time, Rees worked to promote and improve science and science education in America through her membership on numerous committees. To name just a few: the General Sciences Advisory Panel of the Department of Defense, the National Science Foundation Board, Chairman of the Council of Graduate Schools of the United States, and the American Association for the Advancement of Science, becoming the first female President of that organization in 1971. The Mathematical Association of America bestowed its first Award for Distinguished Service to Mathematics on Rees in 1962. In 1983 she received the National Academy of Sciences Public Welfare Medal.

Over the years she wrote many articles, primarily on war research and the ONR, the development of computers, and education.[25] Throughout her career she contributed money to many educational projects, such as the Schafer Prize Fund of the Association for Women in Mathematics. Upon her retirement from the Presidency of CUNY in 1972, she was named Emeritus Professor. Rees continued her involvement in many committees well into the 1980s. Retirement also gave her more time to enjoy and pursue the arts. She was interested in several art forms, but particularly painting. She had spent several vacations, before and after retirement, studying painting in Mexico and Maine. During her Presidency at GSUC, Ray Ring, director of building design and exhibitions at GSUC, presented her with a book of her sketches.

Among Rees's other interests were playing tennis and playing the piano. She was active in the New York community and was a member of All Souls Unitarian Church in Manhattan for many years. In honor of her contribution

---

[24]Some of the women's colleges had female deans.

[25]See the bibliography for a listing of Rees's publications.

to CUNY and GSUC, the Graduate School library was rededicated as the Mina Rees Library in 1985, the first U.S. library dedicated to a mathematician. Mina Rees died at the Mary Manning Walsh Home in Manhattan on October 25, 1997, at the age of 95.

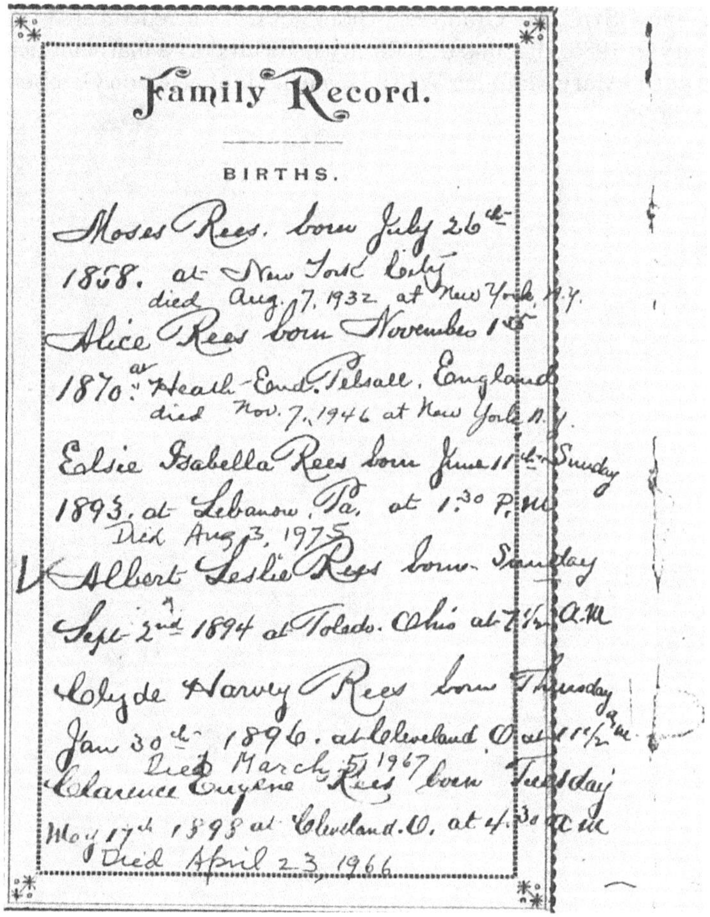

Family Record.

BIRTHS.

Moses Rees. born July 26th
1858. at New York City
died Aug. 7. 1932. at New York, N.Y.
Alice Rees born November 1st
1870 at Heath End. Pelsall, England
died Nov. 7, 1946 at New York N.Y.
Elsie Isabella Rees born June 11th Sunday
1893. at Lebanon, Pa. at 1:30 P.M.
Died Aug. 3. 1975
Albert Leslie Rees born Saturday
Sept. 2nd 1894 at Toledo. Ohio at 7½ A.M.
Clyde Harvey Rees born Thursday
Jan 30th 1896. at Cleveland O. at 11½ a.m.
Died March 5, 1967
Clarence Eugene Rees born Tuesday
May 17th 1898 at Cleveland. O. at 4:30 a.m.
Died April 23, 1966

FIGURE 2.1. Rees family birth record. Courtesy of the Archives of the Graduate Center, CUNY.

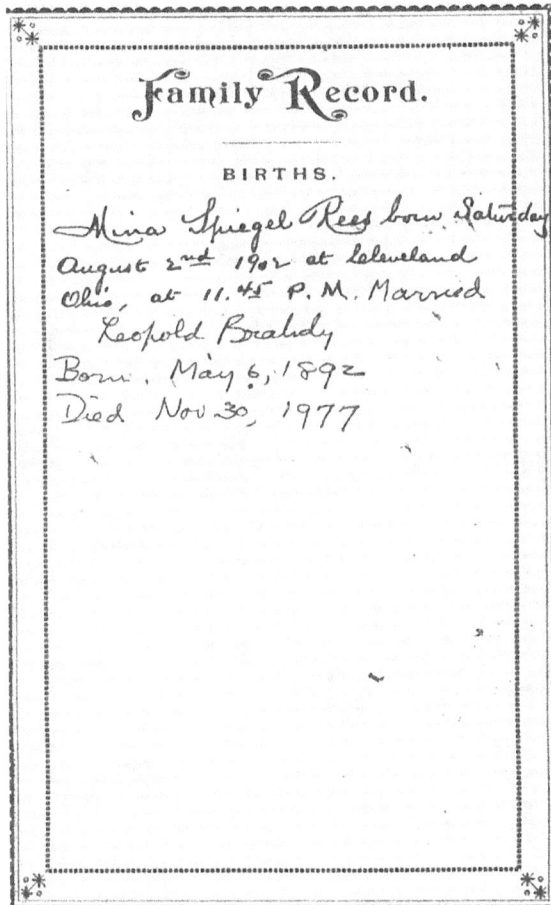

FIGURE 2.2. Rees family birth record.   Courtesy of the Archives of the Graduate Center, CUNY.

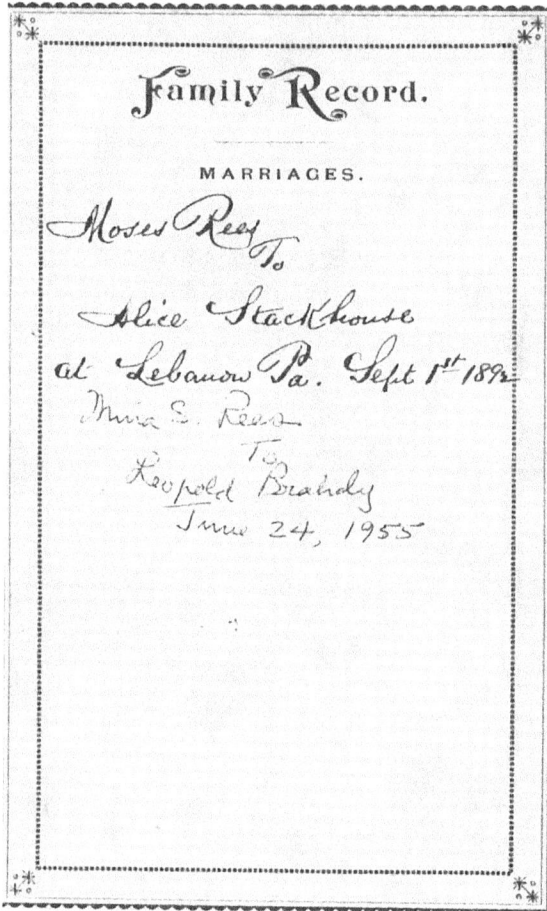

FIGURE 2.3. Rees family marriage record. Courtesy of the Archives of the Graduate Center, CUNY.

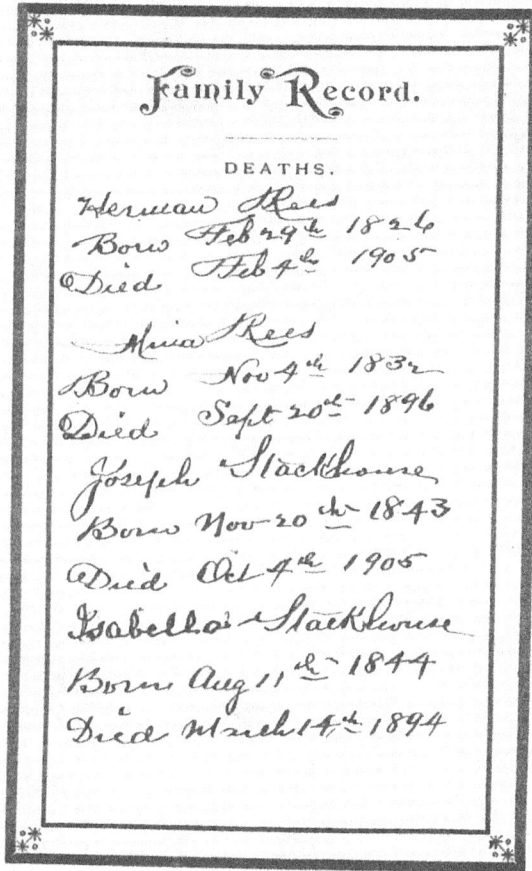

FIGURE 2.4. Rees family death record. Courtesy of the Archives of the Graduate Center, CUNY.

FIGURE 2.5. Sketch by Mina Rees, 1983. Most likely of Burley Island, Maine. Courtesy of the Archives of the Graduate Center, CUNY.

FIGURE 2.6. Sketch by Mina Rees. Location unknown, most likely Maine. Courtesy of the Archives of the Graduate Center, CUNY.

FIGURE 2.7. Sketch by Rees while in Maine, c. mid 1980s. Courtesy of the Archives of the Graduate Center, CUNY

CHAPTER 3

Division Algebras

Mina Rees's doctorial dissertation,[1] *Division Algebras Associated with an Equation Whose Group Has Four Generators,* was completed under the guidance of Leonard Dickson at the University of Chicago in December of 1931. It was published in the *American Journal of Mathematics,* volume 54, 1932.[2]

In looking at Rees's dissertation, two questions come to mind. First, why did Dickson and Rees study associative division algebras? Second, why would we be interested in their work today? The answer to the first question is straightforward. Rees studied Dickson's text *Algebras and Their Arithmetics*[3] while at Columbia, and fell in love with the topic. She decided that Dickson was the man to work with.

> I decided that Dickson was the greatest man in the world and Chicago was undoubtedly the mecca of all algebraists, so without mentioning it to Chicago, I just went out in [1929] and turned up ...I got there only to discover that Dickson was no longer working in algebra. He had given his full attention to number theory a couple of years earlier. I just informed him I wanted abstract algebra, and there just wasn't any work going on in abstract algebra at Chicago while I was there, so I am virtually self-educated. It was the craziest arrangement.[4]

---

[1] This chapter on Dickson's and Reess mathematical research can be skipped in part or in full with little loss to the narrative of Rees life and work.

[2] The work published in the *Journal of Mathematics* is her actual dissertation except with the omission of one table.

[3] [**12**].

[4] [**39**].

The above quote gives an interesting look at Rees's personality, and possibly an indication of her mode of operation at ONR and CUNY later on. Why did she go to Chicago without contacting Dickson or the University? Was she worried about being accepted to the University? That does not seem to fit, given her background. Did she decide that this was what she wanted, and that just going was the most efficient or effective action? In either case, her strategy was successful, not to mention bold.

It is also curious that Rees says there was no algebra being done at Chicago then. Dickson's most influential paper on division algebras appeared in 1930. He continued to write papers and give talks on division algebras into the 1930s. Also, Adrian Albert received his Ph.D. under Dickson in 1928 with a dissertation on division algebras. He returned to the University in January 1931 as a faculty member, the year Rees finished at Chicago.

Before considering Dickson's work in division algebras, a brief history of such algebras will be given. The theory of algebras had three fairly distinct lines of development, Britain (Hamilton, Cayley), Germany (Noether, R. Brauer, Hasse), and the United States (Peirce, Dickson, Wedderburn, Albert). This brief history will center on work done in the United States, and particularly, that which relates to the work of Rees. First a short history of the subject will be given, then the three papers of Dickson's that are relevant to Rees's work will be discussed, finally Rees's Ph.D. dissertation will be explored.

First, what is an associative division algebra? In modern terminology, an algebra A, is a vector space over a commutative field F, such that the left and right distributive laws hold for A, A being a ring; A is an associative algebra if multiplication is associative in A, and a division algebra if A is a division ring. For our discussion "associative division algebra" will mean a finite dimensional associative division algebra, hence, a finite dimensional vector space with multiplication over a field.

3.1. A Brief History of Algebras

In 1834 Sir William Rowan Hamilton discovered the *quaternions,* a non-commutative extension of the complex numbers in which there are three elements i, j, and k such that

$$i^2 = j^2 = k^2 = ijk = -1$$
$$ij = -k, \quad ji = k$$
$$ik = j, \quad ki = -j$$
$$jk = -i, \quad kj = i.$$

The quaternions form a vector space over the reals with basis $\{1, i, j, k\}$. Hamilton then extended the quaternions to have base field the complex numbers, dubbing this system the *biquaternions.* The study of quaternions, or hypercomplex numbers, became very popular in Britain. John Graves developed the octaves or octonions in 1844, an eight dimensional analogue of quaternions in which neither commutativity nor associativity holds. Hamilton's quaternions became a vector space in light of work done by Josiah Gibbs and Oliver Heaviside in the 1880s on vector spaces.

By 1848, interest in the subject had reached the United States. Benjamin Peirce became interested in generalizing the quaternions over the complex numbers, referring to them as linear algebras. Peirce made many contributions to the field, including the idea of nilpotent elements (an element a is nilpotent if $\exists n$ s.t. $a^n = 0$), and idempotent elements (an element a is idempotent if $a^2 = a$). Probably his biggest contribution is the Peirce Decomposition Theorem. Peirce also cataloged all complex associative linear algebras of dimension up to six.[5]

Peirce's paper, "Linear Associative Algebras," was published privately in 1870, and posthumously in 1881 by his son C. S. Peirce, who included an important finding of his own: any associative algebra can be represented as an algebra of matrices. Later he proved that the only associative division algebras of finite dimension over the reals were the reals, the complex numbers, and the quaternions. This was also proven independently by Georg Frobenius.

---

[5][**31**].

Due to the influence of the Peirces' work, cataloging linear associative algebras became a primary focus in the United States around the turn of the century. About 1904 Leonard Dickson at the University of Chicago and Joseph Wedderburn in Scotland became involved in this effort, in particular, the description of division algebras. Wedderburn was influenced by the work of Frobenius and Elie Cartan on hyper-complex numbers. In 1926, Dickson stated that the most important problem in linear algebras was the determination of all division algebras.

Leonard Eugene Dickson was born January 22, 1874, in Independence, Iowa. Dickson's father, Campbell Dickson, combined the pioneering spirit of the west with the entraprenurial drive of the east and moved his family west. Dickson's mother held a college degree and taught school before raising a family. These two strong personalities would exert a strong influence on Dickson throughout his life. He grew up in Texas and attended the University of Texas, obtaining his bachelor's degree in mathematics in 1893 and his master's in 1894. While at Texas, he studied Euclidean and non-Euclidean geometry. He then attended the fledgling University of Chicago, obtaining his Ph.D. under E. H. Moore in 1896.[6] His dissertation was on linear groups. Fenster notes that "Dickson joined the first generation of aspiring American mathematicians who chose to pursue their doctorates at institutions in their home country."[7]

He then went to Europe to broaden his mathematical background by working with Lie, Picard, Hermite and Jordan. Returning to the United States, he taught at the University of California at Berkeley for two years. He then accepted a three-year appointment to the University of Texas at Austin, only to leave after one year to go to the University of Chicago in 1900 to work again with E. H. Moore.[8] His areas of study were primarily algebras, group theory, invariant theory, and number theory. In 1911, while still at the height of his research, Dickson started work on what may be one of his most well known works, his three-volume (a forth volume remained unpublished) *History of the Theory of Numbers* (1919, 1920, 1923). Dickson retired from Chicago in 1939, after 40 years on the faculty. He died January 17, 1954, in Harlingen, Texas.

---

[6]The original mathematics department at Chicago, from whom Dickson took classes, consisted of E. H. Moore, Oskar Bolza and Heinrich Maschke.

[7][**21**, p. 55].

[8]*Ibid.*, p. 56.

Dickson supervised sixty-seven doctoral students, eighteen of whom were women. E. H. Moore praised Dickson's work by saying that he was, "one of the very strongest research men in mathematics in the whole country."[9]

Historian of mathematics William Duren had these thoughts on Dickson.

> Dickson was not much of a teacher. I think his students learned from him by emulating him as a research mathematician more than being taught by him. Moreover, he took them to the frontiers of research, for the subject matter of his courses was usually new mathematics in the making.[10]

Dickson began his work in algebras about 1903 when he introduced the concept of a linear division algebra. By 1905 he was considering finite division algebras in which commutativity or distributivity failed. He also extended the definition from the real and complex numbers to division algebras of arbitrary fields.[11] Also in 1905 Dickson and Wedderburn proved that any finite division algebra is a field. Wedderburn also showed that the only finite associative division algebras over the reals were finite Galois fields.

Dickson would work on division algebras for twenty years, continuing to refine the definition of a division algebra, giving three more definitions. Fenster states that,[12]

> [a]s he re-evaluated this concept in light of his mathematical researches (especially in division algebras and, later, in the arithmetic of algebras), Dickson emphasized the linear independence of the basis elements with respect to the field of coefficients. This allowed for a more far-reaching definition, one which included the properties of division and associativity and, in 1923, put forth a definition of an algebra as we know it today. ... In a broad sense, he lent a certain consistency to the

---

[9]*Ibid.*, p. 57.
[10][**20**, p. 14].
[11][**21**, p. 56].
[12]*Ibid.*, p. 67

foundations of the theory, pursuing research on this aspect of the field for over two decades.

1907 saw three important results by Wedderburn:

1. Any algebra is the sum of a nilpotent algebra and a semi-simple algebra.
2. Any non-simple semi-simple algebra is the direct sum of simple algebras.
3. Any simple algebra is the direct product of a division algebra and a full matrix algebra.

A semi-simple algebra is one that contains no nilpotent two-sided ideal, and a simple algebra is an algebra with no two-sided ideal. In these results, the base field is assumed to have characteristic zero. Of particular importance to the remainder of our discussion, Dickson suggested that the term algebra be used in place of linear associative algebra.

These results yield Wedderburn's Structure Theorem: that any algebra over a field is simple if and only if it is the direct product (commutative tensor product) of a matrix algebra and a division algebra.

Up to 1905, the only known division algebras were the quaternions and fields. In 1905, Dickson constructed new non-commutative division algebras over arbitrary fields. This was a major finding. He showed that to every field with a cyclic extension, there can be associated a division algebra. In modern terminology, these are the cyclic division algebras. Dickson's two textbooks and numerous articles on division algebras up to 1930 greatly influenced the path of research in this field in the United States. More detail about Dickson's findings is provided in the next section.

By 1924 Wedderburn was looking at infinite division algebras, while by the 1930s Dickson was investigating non-associative algebras, and had for the most part turned to the study of number theory. The next person of interest to this narrative is Adrian Albert (1905–1972). Albert studied division algebras under Dickson at Chicago, receiving his doctorate in 1928. Albert did much work in constructing normal (central) division algebras of low degree. He showed that every division algebra over its center of degree four is a crossed product. An algebra is said to be a crossed product if it is normal simple of dimension n^2

over its field $\mathcal{F}$ of scalars, and has a subfield which is normal of degree n over $\mathcal{F}$. A normal algebra is one in which its base field is its center (it is an algebra over its center, or a centered algebra). As Wedderburn noted in 1921, if A is a division algebra, then the center of A can be taken to be the base field of A.

By the late 1920s, Albert was applying central simple algebras to the study of Riemann surfaces, in particular the problem of commutative Riemann matrices. Albert's biggest result was that every central division algebra over an algebraic number field of finite degree is cyclic, but this result was published in 1932 by Noether, R. Brauer, and Hasse just ahead of him.[13] Much of Albert's earlier work on algebras appears in his text *Structures of Algebra* (1939).

Emmy Noether (1882–1935) was one of the most influential algebraists of the 20th century. One of her greatest impacts on algebra was in defining algebras in terms of set theory, known as 'set theoretic algebra'. Her analysis was based on defining the properties of an algebraic structure by defining the properties of its substructure by exploring subdomains such as the subgroups, normal subgroups or subrings of the algebra. By looking past the elements of the algebra and looking at the relationship between the elements, she was able to reach more general yet at the same time more insightful, deep and far reaching conclusions. In particular, her use of homomorphisms and isomorphisms gave her the power to generalize much more quickly than the elemental approach taken by Dickson and Rees. This approach is summarized in Noethers Principle: To base all of algebra, so far as possible, on considerations of isomorphisms.[14] Noether's school of thought quickly permeated much of mathematics. For example, Brouwer's ideas on mappings and spaces combined with Noethers ideas on algebra, in particular group theory, were applied to topology, effectively making it algebraic. Alexandroff and Hopf lay this paradigm shift in topology firmly at Noethers feet, "The tendency to strict algebraization of topology on group theoretic foundations . . . goes back entirely to Emmy Noether."[15]

The study of division algebras is still active today. Fein and Schafer showed in 1976 that algebras can be uniquely represented as crossed products. In the

---

[13][**31**].

[14][**37**, p. 220].

[15]*Ibid.*, p. 215

1980s, Amitsur showed that there are non-crossed product algebras of dimension n, if n is divisible by eight or by the square of an odd prime.[16]

3.2. Why Study Division Algebras?

The study of algebras is very interesting from an historical standpoint. Emmy Noether and her colleagues were also working on algebras during the late 1920s, but from a different perspective. Dickson's approach was constructive, using polynomials and their roots to generate his division algebras. While Dickson did consider the Galois group, he did not use Galois theory in his constructions. On the other hand, Noether's approach was much more set theoretic and used Galois theory to effect. Her approach, as noted above, was the more general and generalizable, since Dickson's approach required numerous relations to form each new extension. The Noether school took a global view, looking for generalized structures of algebras, whereas Dickson took an elemental approach that looked at the behavior of the elements of the algebra. Noether's method quickly became standard while Dickson's became obsolete at about the time Rees wrote her dissertation. Thus our study gives a nice example of independent research being run concurrently, developing in different ways, and having different impacts.

Although Dickson's work and Rees's work may have been superseded by the work of others, it did lead to other areas of research and impacted other fields. At the time Rees became interested in division algebras, the American school, with Dickson in the forefront, was occupied with classifying division algebras. In particular, division algebras play an important role in matrix and ring theory. Dickson applied algebras to number theory and Diophantine equations. Albert applied them to algebraic geometry, and developed the basic theory of finite Jordan algebras over fields of characteristics not equal to two. Jordan, Wigner, and von Neumann would apply algebras to quantum mechanics. Results of Amitar and Saltman in 1978 on generic abelian crossed products actually rediscovered old results of Dickson.

---

[16][**35**, p. 4].

3.3. Dickson and Rees

As already mentioned, Rees arrived in Chicago in the summer of 1929. She received her doctorate at the end of 1931, after only five semesters at the university.[17] Della Fenster observed that,

> Dickson's teaching, it seems, reflected his lifelong goal to become the most distinguished mathematician possible. He spent his mathematical life at the cutting edge of the field, and he wanted his students to do the same. Since students who could not meet his standards also could not serve his purposes best, Dickson had a sudden death trial for his prospective doctoral students: he assigned a preliminary problem which was shorter than a dissertation problem, and if the student could solve it in three months, Dickson would agree to oversee the graduate student's work.[18]

Dickson was a top researcher in the United States, having been voted one of the top seven mathematicians in the county in the 1910 edition of *Men of Mathematics*. As such, he was a very sought after doctorial advisor. There is no evidence that Rees was given such a problem, and given the short amount of time Rees was in Chicago, it is unlikely she was given one. It appears she simply convinced Dickson to give her a problem in division algebras. But as we will see shortly, this may have been beneficial to them both. However, she may have been aided in her case. Still being a member of the Hunter College faculty, Rees was invited to represent Hunter at the the University of Chicago's inauguration dinner for its new president, a very singular honor. Soon after this Dickson invited Rees to become his PhD student.[19]

Since Rees was adamant about writing a dissertation in division algebras, she most likely began by reading his work in the field. Dickson's publications

---

[17]See Appendix A for Rees's course work at the University of Chicago. The permanent mathematics faculty at Chicago while Rees attended was, in addition to Dickson: Mayme Logsdon, Ernest Lane, Eliakim Hastings Moor, Raymond Barnard, Lawrence Graves, Gilbert Bliss, and Arthur Lunn.[88]

[18][20, p. 14].

[19][30].

on division algebras in the first two decades of the twentieth century were cumulative. Each paper built on the foundation set by the previous one. Rees's dissertation serves as a final and culminating paper in this series.

In general, given a normal extension $\mathcal{F}(\alpha)$ of a field $\mathcal{F}$, Dickson constructs a division algebra over $\mathcal{F}$. We will describe Dickson's work in terms of field extensions and automorphisms belonging to the Galois group, but it is important to keep in mind that Dickson's approach was not as abstract as this might imply. He used polynomials that permuted the roots, and numerous relations on his basis elements to construct associative division algebras. In order to present Dickson's and Rees's work as concisely as possible, a more modern approach, which will be more familiar to the reader, will be being taken. The presentation here relies more heavily on group theoretic notions, more along the lines of Noether. But every attempt has been made to maintain the feel of the Dickson and Rees work.

To understand Rees's work, we will look at three papers of Dickson's on associative division algebras, published in 1914, 1926, and 1930. In general, these papers start with the simplest case, in which the Galois group is cyclic, then move on to the abelian case, and finally the solvable case. The constructions of the algebras progress from concrete to general. Rees's work follows from Dickson's most general paper, that of 1930.

3.4. Dickson 1914

The first paper that is relevant to our discussion of Rees's dissertation is Dickson's 1914 paper, "Linear Associative Algebras and Abelian Equations."[20] He starts this paper by providing the details for an announcement he made in the *Transactions of the American Mathematical Society* in 1906: the construction of a cyclic division algebra $\mathcal{A}$ with q^2 basis elements over a field $\mathcal{F}$.

In modern terminology, the cyclic case is as follows. Let $f(x) = 0$ be an irreducible polynomial over $\mathcal{F}$ of degree q with primitive root α, and suppose

---

[20]All three of the Dickson papers discussed were published in the *Transactions of the American Mathematics Society.*

that the Galois group $\mathcal{G}$ of $f(x)$ is cyclic. That is

$$\mathcal{G} = \{I, \Theta, \Theta^2, \ldots, \Theta^{q-1}\},$$

for some $\Theta \in Aut(f)$. Hence

$$\Theta(\alpha), \Theta^2(\alpha), \ldots, \Theta^{q-1}(\alpha), \Theta^q(\alpha) = \alpha$$

are the q distinct roots of $f(x)$, where $\Theta^k(\alpha)$ denotes composition. $\mathcal{F}(\alpha)$ is the corresponding normal field extension of $\mathcal{F}$ of degree q. ($\mathcal{F}(\alpha)$ is a splitting field for which $\mid \mathcal{F}(\alpha) : \mathcal{F} \mid = q$.)

This special case in which the Galois group is cyclic[21] motivated Dickson to look for methods of determining what field extensions are division algebras.

The purpose of Dickson's 1914 paper is to develop a method to construct all division algebras, $\mathcal{A}$, of low dimension with the following properties:

 i) if $x \in \mathcal{A}$ such that $x\alpha = \alpha x$, then $x \in \mathcal{F}(a)$
 ii) $\exists j \in \mathcal{A}\backslash\mathcal{F}(\alpha)$ such that the following properties hold:
   1) $j^q = g \in \mathcal{F}(\alpha)$, where q is the least such exponent
   2) $j\alpha = \Theta(\alpha)j$.

The existence of the root α, and the conditions above tell us that $\mathcal{A}$ is a crossed product;[22] $\mathcal{A}$ will be of degree q^2 over $\mathcal{F}$ with basis elements

$$\{\alpha^r j^s \mid r, s = 0, 1, 2, \ldots, q-1\}.$$

Dickson proves that $\mathcal{A}$ will be a division algebra if and only if g is not the norm of any element of $\mathcal{F}(\alpha)$. He next gives an example in which $q = 2$. He refers to this case as the generalized quaternions.

Dickson concludes his 1914 paper with an example of an algebra with sixteen basis elements (i.e. $q = 4$). Let $f(x)$ be an irreducible polynomial over some field $\mathcal{F}$ with root α, such that $\mathcal{F}(\alpha)$ is a normal field extension. Then $f(x)$ will be (after a translation if necessary) of the form

$$x^4 + 2Bx^2 + C = 0$$

---

[21]A group is cyclic if it is generated by a single element.
[22]Recall that a crossed product algebra is an algebra that is normal simple over its field of scalars, of degree n^2, with center its base field.

where B and C are in $\mathcal{F}$. Since $\mathcal{F}(\alpha)$ is normal, then α is a primitive element. $1, \alpha, \alpha^2, \alpha^3$ will form the basis for the splitting field, which will be $\mathcal{F}(\alpha)$, and Θ is completely determined by its action on α. Extend $\mathcal{F}$ to a splitting field by adjoining $\alpha, -\alpha, \beta,$ and $-\beta$. Let $\Theta \in \mathcal{G}$. Since the Galois group has four elements, it must be either the cyclic group of order four or the four group.

It is not hard to see that the Galois group is cyclic if and only if the coefficient C in $f(x)$ is not the square of an element in $\mathcal{F}$. This causes $f(x)$ to have the form
$$x^4 - 2bdvx^2 + vd^2 = 0.$$
In this case, $\mathcal{G} = \{I, \Theta, \Theta^2, \Theta^3\}$ where $\Theta^4 = I$. $\mathcal{G}$ has only one automorphism of order two, namely Θ^2. So $\Theta^2(\alpha) = -\alpha$. Then $\Theta(\alpha) = \pm\beta$. For each $g \neq 0$ in $\mathcal{F}$, there is an associative algebra $\mathcal{A}$, such that $j\alpha = \Theta(\alpha)$, $j^4 = g$, and $\{\alpha^m j^n \mid m, n = 0, 1, 2, 3\}$ form the basis.

The case in which the Galois group is the four group arises when C is the square of some element of $\mathcal{F}$. Then $f(x)$ will be of the form
$$x^4 - 2dax^2 + d^2 = 0.$$
Then the splitting field will be $\mathcal{F}(\sqrt{w}, \sqrt{t})$ for some $w, t \in \mathcal{F}$, and $\mathcal{G}$ will have two elements of order two, Θ and Φ. For each g and h in $\mathcal{F}$, we can form an algebra with the following properties:

    i) $j^2 = g$ and $k^2 = h$ for $j, k \in \mathcal{A}/\mathcal{F}(\sqrt{w}, \sqrt{t})$
    ii) $jz = \Theta(z)j$, $kz = \Phi(z)k$, and $kj = \alpha jk$.

Note, $\Theta(g) = g$ and $\Phi(h) = h$, since $g, h \in \mathcal{F}$.

Using the second set of relations above, we can calculate kj^2 and k^2j by assuming associativity.[23] This leads to the two conditions needed for $\mathcal{A}$ to be associative:

    iii) $\Phi(g) = \alpha\Theta(\alpha)g$
    iv) $h = \alpha\Phi(\alpha)\Theta(h)$.

---

[23]For example, (iv) is found by the following calculation: $k^2j = kkj = \Phi(\alpha)kjk = \Phi(\alpha)\alpha jk^2 = \alpha\Phi(\alpha)jk^2 = \alpha\Phi(\alpha)jh = \alpha\Phi(\alpha)\Theta(h)j$. So $k^2j = hj$. Therefore, $hj = \alpha\Phi(\alpha)\Theta(h)j$. And thus, $h = \alpha\Phi(\alpha)\Theta(h)$.

The 1914 paper is of interest since it shows how concrete Dickson's methods were. He explores only the cases of dimension four and sixteen. He also shows that the eight-dimensional case is impossible. The nine-dimensional case is dealt with in his 1923 text.[24] His study of the example with dimension sixteen prepared him to study more general cases in which the group is abelian. This leads to the work presented in his next paper.

3.5. Dickson 1926

"New Division Algebras" is a transitional paper in Dickson's work in associative algebras. The 1914 paper is very concrete in its approach. While by 1930, Dickson had developed a more general method within the framework of crossed products that handles more general cases of the Galois group. In the 1926 paper he looks at both the abelian and solvable case, and lays the framework for the 1930 paper. The opening paragraphs of this paper show that Dickson realized that he was breaking new ground.

> The chief outstanding problem in the theory of linear algebras (or hypercomplex numbers) is the determination of all division algebras. We shall add here very greatly to the present meager knowledge of them, since we shall show how to construct one or more types of division algebras of order n^2 corresponding to every solvable group of order n.
> While it was long thought that the theory of continuous groups furnishes an important tool for the study of linear algebras, the reverse position is now taken. But this memoir shows how vital a role the theory of finite groups plays in the theory of division algebras.[25]

Dickson begins his 1926 paper with the announcement of the construction of division algebras in which the Galois group is abelian or doubly cyclic, with two independent generators. This is a generalization of the example at the end of his previous paper,[26] in which the Galois group is the four-group.

---

[24]*Algebras and Their Arithmetics*, [**12**]
[25][**13**].
[26]See end of previous section.

His construction is as follows.[27] Let $\mathcal{A}$ be an algebra over a field $\mathcal{F}$ as described above, with degree qp and primitive element α. The normal field extension $\mathcal{F}(\alpha)$ will be of order qp, and have Galois group

$$\mathcal{G} = \{\Theta^r \Phi^s \mid 0 \leq r < q, 0 \leq s < p\}.$$

For any non-zero elements g_1 and g_2 in $\mathcal{F}(\alpha)$, the algebra $\mathcal{A}$ is formed by adjoining elements j and k not in $\mathcal{F}(\alpha)$ such that the following conditions hold:

$$\begin{aligned}
j^q &= g_1 \\
k^p &= g_2 \\
j\alpha &= \Theta(\alpha)j \\
k\alpha &= \Phi(\alpha)k \\
kj &= \alpha jk.
\end{aligned} \tag{3.1}$$

Thus,

$$\{j^r k^s \mid 0 \leq r < q, 0 \leq s < p\}$$

form a basis for $\mathcal{A}$ over $\mathcal{F}(\alpha)$ of order qp. Hence,

$$\{\alpha^t j^r k^s \mid 0 \leq t < qp, 0 \leq r < q, 0 \leq s < p\}$$

form a basis for $\mathcal{A}$ over $\mathcal{F}$. $\mathcal{A}$ will be associative under the following conditions:

$$\Theta(g_1) = g_1, \Phi(g_2) = g_2 \tag{3.2}$$

$$\Phi(g_1) = \alpha\Theta(\alpha)\Theta^2(\alpha)\cdots\Theta^{q-1}(\alpha)g_1 \tag{3.3}$$

$$g_2 = \alpha\Phi(\alpha)\Phi^2(\alpha)\cdots\Phi^{p-1}(\alpha)g_2 \tag{3.4}$$

The conditions 3.2, 3.3, and 3.4 are found by calculating kj^q and $k^p j$ using the conditions 3.1 above.

It is important to understand Dickson's approach before considering his actual method for developing division algebras. He assumes that he has a finite dimensional algebra, $\mathcal{A}$, and then determines the conditions necessary for it to be an associative division algebra. He assumes that $\mathcal{A}$ has a maximal subfield that is also a normal extension of $\mathcal{F}$. Then a basis can be chosen that will give the crossed product algebra. This assures that there is a primitive[28] element

---

[27]One of the difficulties of reading the papers of Dickson and Rees is that Dickson, and Rees to some extent, changed notation from construction to construction, and paper to paper. For clarity, the same notation will be maintained throughout the discussion of all four papers.

[28]An element is primitive if it is invertible.

α and thus an irreducible polynomial $f(x)$ over $\mathcal{F}$ so that $f(\alpha) = 0$, and the dimension of $\mathcal{A}$ will be the square of the degree of $f(x)$.

The second part of this 1926 paper deals with the general (non-abelian) case. From this point on Dickson is interested in finding all division algebras that are crossed products. This work was started by Wedderburn, and will provide the framework for Dickson's construction of division algebras.

Thus he wants to find all division algebras having the following properties as stated by Dickson:[29]

  (I) $\mathcal{A}$ is of order n^2;
 (II) $\mathcal{A}$ contains an element[30] α which satisfies an equation $f(x) = 0$ of degree n, irreducible in $\mathcal{F}$;
(III) The only elements of $\mathcal{A}$ which are commutative with α are polynomials in α, with coefficients in $\mathcal{F}$;
(IV) The roots of $f(x) = 0$ are all rational functions $\Theta_r(\alpha)$ of α, with coefficients in $\mathcal{F}$.

Let α be a primitive element of order n, and $\mathcal{F}$ the base field. Take the Galois group of the normal extension $\mathcal{F}(\alpha)$ over $\mathcal{F}$ to be

$$\mathcal{G} = \{I, \Theta_1, \Theta_2, \ldots, \Theta_{n-1}, \Phi_1, \Phi_2, \ldots, \Phi_{n-1}\}$$

Construct $\mathcal{A}$ as before, such that for each Θ_i, $0 < i < n$, there is a j_i in $\mathcal{A}\backslash\mathcal{F}(\alpha)$ such that:

$$j_i\alpha = \Theta_i(\alpha)j_i$$

Then $\{\alpha^m j_i \mid m, i < n\}$ is a basis for $\mathcal{A}$ over $\mathcal{F}$, $(j_0 = 1)$. In the general case, the algebra will be a division algebra if each power less than n of g_i for each i, is not the norm of any element in the extension field. Dickson states that this was proven by Wedderburn in 1914.[31]

The general case can be summarized by the following theorem.

---

[29][**13**, p. 208–209].

[30]Dickson used i for the root, but this is too easily confused with the imaginary use of i. For clarity, we will continue to use α as he did in the earlier paper.

[31][**13**, p. 228].

THEOREM. *Let $\mathcal{A}$ be an algebra with multiplicative identity e, $\mathcal{F}$ the base field of $\mathcal{A}$ as a subfield of $\mathcal{A}$, $\mathcal{F}(\alpha)$ a normal extension of $\mathcal{F}$ and a subfield of $\mathcal{A}$,*

$$\mathcal{G} = \{I, \Theta_1, \ldots, \Theta_{n-1}\}$$

the Galois group of $\mathcal{F}(\alpha)$ over $\mathcal{F}$. Then there exists

$$j_0 = e, j_1, \ldots, j_{n-1} \in \mathcal{A} \backslash \mathcal{F}(\alpha)$$

such that

    (a) $\mathcal{A} = \mathcal{F}(\alpha)j_0 + \mathcal{F}(\alpha)j_1 + \cdots + \mathcal{F}(\alpha)j_{n-1}$
    (b) $j_i\beta = \Theta_i(\beta)j_i, \forall \beta \in \mathcal{F}(\alpha), 1 < i < n$
    (c) $\forall \Theta_i, \Theta_k \in \mathcal{G}, \exists c_{i,k} \in \mathcal{F}(\alpha)$ *such that* $j_i j_k = c_{i,k} j_{ik}$, *where j_{ik} is the element in $\mathcal{A}$ corresponding to $\Theta_k \circ \Theta_i$.*

Since $\mathcal{A}$ is to be a division algebra, $j_i j_k \neq 0$ implies $c_{i,k} \neq 0$. The conditions for $\mathcal{A}$ to be associative are given explicitly by Adrian Albert as:[32]

$$\forall \Theta_i, \Theta_k, \Theta_l \in \mathcal{G}, c_{i,k}c_{li,k} = \Theta_l(c_{i,k})c_{l,ik}$$

and using (b) and (c) above.

From our discussion of Dickson's 1914 paper, the center of $\mathcal{A}$ is $\mathcal{F}$. Thus the algebra $\mathcal{A}$ constructed is a normal simple associative division algebra, and is a crossed product algebra.

Dickson next passes to the case where $\mathcal{G}$ is an arbitrary abelian group. If the degree of the irreducible polynomial is n, then the algebra $\mathcal{A}$ will have order n^2. Dickson uses the important fact that every finite abelian group is a direct product of cyclic groups. He thus avoids the possibility of $\Phi^a = \Theta^b$. The general abelian case can be explained by generalizing the previous case to any number of independent cyclic generators.

In the last sections of the 1926 paper, Dickson returns to topics touched on earlier in the paper. First he gives a few concrete examples of the abelian case with two generators, in particular, $p = q = 2$ and $p = q = 3$. Then he gives some partial results for the solvable case. These discussions are even more constructive than the previous discussions. To handle these cases, Dickson relies

---

[32][1].

heavily on the matrix representation of an algebra. Since the general crossed product method, started here and developed fully in his 1930 paper, makes the approach via matrix methods obsolete, we will not consider the matrix representation here.

3.6. Dickson 1930

Dickson's 1930 paper, "Construction of Division Algebras," is his most important work in division algebras. It provides his most general and least cumbersome method for constructing crossed product algebras. He stated that,

> The main outstanding problem in the theory of linear associative algebras is the determination of all division algebras. We shall make a noteworthy simplification of the theory of the construction of a type of algebras Γ which includes all known division algebras. The simplification is so great that it would require a hundred pages to obtain our results by the best earlier method.[33]

He starts by providing the details for the solvable case on two generators. He then continues the work of his 1926 paper by generalizing the case in which the Galois group is abelian of degree $d = p_1 p_2 \cdots p_n$, made up of n cyclic groups of order p_i, $i = 1, 2, \ldots n$. He then goes on to the case in which the Galois group is solvable with three generators. Dickson's order of presentation (solvable with two generators, general abelian, solvable with three generators) may seem strange. Dickson uses the two-solvable case to illustrate the use of what we will call his Extension Theorem. The general abelian case is not suitable for this purpose because it is too general. It requires a proof by induction on n, where the Extension Theorem is used for the inductive step. Dickson does not explicitly point this out, nor where the theorem is used, making the reading of this part of his paper difficult. We will look closely at the solvable case with three generators, since Rees assumes this case in her dissertation and builds her work on this case inductively. Dickson remarked, "We postpone extensions to non-Abelian equations whose roots have four or more generators."[34] The

---

[33][**14**, p. 319].
[34]*Ibid.*

case of four generators is Rees's dissertation. As Dickson's quote at the start
of this section shows, as time went on, the methods for developing associative
division algebras became more succinct. But even Dickson's 1930 paper and
Rees's dissertation require long calculations. This is due to the fact that they
are not working fully in Galois theory, as Noether did. The case of more than
four generators was not undertaken, with good reason.

To begin our study of Dickson's method of constructing algebras, we will
give a few definitions and theorems. The expression 'solvable group with n
generators' is taken to mean the following:

DEFINITION. *Let $\mathcal{G}$ be a group generated by automorphisms $\Psi_1, \Psi_2, \ldots \Psi_n$
of order $p_1, p_2, \ldots, p_n$ respectively. Then we will say that $\mathcal{G}$ is solvable with n
generators if there is a normal sequence of subgroups.*

$$\{I\} = \mathcal{G}_1 \lhd \mathcal{G}_2 \lhd \cdots \lhd \mathcal{G}_n = \mathcal{G}$$

where

$$\mathcal{G}_{i+1} = \bigcup_{r < p_{i+1}} \mathcal{G}_i(\Psi_{i+1})^r$$

and each factor group $\mathcal{G}_{i+1}/\mathcal{G}_i$ is cyclic.

In overview, Dickson's new construction is to build the desired algebra inductively. This is done by constructing intermediate algebras corresponding to
the adjoining of one generator at a time to the Galois group. This is done by
using the following two theorems. The first, the Extension Theorem, or Theorem 1 of Dickson's 1930 paper, is used to extend algebras. The second, which
will be referred to as the Multiplicative Extension Theorem, is a clarification of
the process Dickson used in the 1930 paper to extend automorphisms, and is
not explicitly stated by Dickson.[35] Both are now stated in modern terms.

EXTENSION THEOREM. *Let Σ be an algebra, with multiplicative identity e,
over a field $\mathcal{F}$ of scalars. Let τ be an automorphism of Σ leaving every scalar
fixed. Let γ be an element of Σ, and p an integer such that:*

    i) $\tau(\gamma) = \gamma$
    ii) $\gamma z = \tau^p(z)\gamma, \forall z \in \Sigma$

---

[35]These extension methods only apply to the solvable case.

Then there exists an extension Γ of Σ, generated by an element J such that:

iii) every element of Γ is uniquely of the form $z_0 e + z_1 J + \cdots + z_{p-1} J^{p-1}$, $z_i \in \Sigma$, for $i < p$

iv) $J^p = \gamma$

v) $Jz = \tau(z)J, \ \forall z \in \Sigma$

We must remember that Dickson does not work in the framework of automorphisms. He uses polynomials that permute the roots of the irreducible equation. His use of polynomials means that any calculations must be worked through for each polynomial, resulting in numerous long and unwieldy conditions. Our use of automorphisms greatly reduces the number of calculations and conditions needed, since we only need to state the condition for the variable used. Note that the abstract viewpoint of Noether allows us to extract this theorem from Dickson's work, and allows us to understand the work of Dickson and Rees in a more modern light.

MULTIPLICATIVE EXTENSION THEOREM.[36] *Let $\mathcal{G}$ be a group with normal subgroup $\mathcal{H}$ such that $\mathcal{G}/\mathcal{H}$ is cyclic of order q, specifically*

$$\mathcal{G}/\mathcal{H} = \{\mathcal{H}, \mathcal{H}b, \ldots, \mathcal{H}b^{q-1}\}$$

where $b^q \in \mathcal{H}$. Let Φ be a homomorphism from $\mathcal{H}$ into some group $\mathcal{G}'$, and suppose b' is an element of $\mathcal{G}'$ such that

i) $(b')^q = \Phi(b^q)$

ii) $b'\Phi(\omega) = \Phi(b\omega b^{-1})b', \ \forall \omega \in \mathcal{H}$

Then there is a homomorphism τ from $\mathcal{G}$ to $\mathcal{G}'$ which coincides with Φ on $\mathcal{H}$ and $\tau(b) = b'$.

These two theorems will supply the conditions of associativity for our constructed algebras, and allows us to count and list them.

---

[36]This theorem was suggested to the author by her dissertation advisor, William Howard, at the University of Illinois at Chicago.

To illustrate the use of these two theorems, we will rework Dickson's result for the solvable case with three generators.[37] To construct our algebra Γ, we start as in the previous sections. Let α be primitive element of $\mathcal{F}$ of degree qps, that provides a normal field extension. Let the Galois group $\mathcal{G}$ be solvable, with qps elements.

Stage 1: Construction of algebra Γ_1

Construct the normal field extension $\mathcal{F}(\alpha)$, call it $\mathcal{F}_1$. Then $\{\alpha^i \mid i < qps\}$ is a basis for $\mathcal{F}_1$. Let

$$\mathcal{G}_1 = \{I, \Theta, \Theta^2, \ldots, \Theta^{q-1}\}$$

where Θ is an element of $\mathcal{G}$ of order q. For each g_1 in $\mathcal{F}_1$, there is a j not in $\mathcal{F}_1$ such that $j^q = g_1 1$ and $\Theta(g_1) = g_1$. Using the Extension Theorem, we construct algebra

$$\Gamma_1 = \mathcal{F}_1 + \mathcal{F}_1 j + \cdots + \mathcal{F}_1 j^{q-1}$$

where $\{j^r \mid r < q\}$ is the basis for Γ_1 over $\mathcal{F}_1$.

Stage 2: Construction of algebra Γ_2

Let Φ be an automorphism from $\mathcal{G}\backslash\mathcal{G}_1$ of order p. Since $\mathcal{G}$ is solvable, let $\mathcal{G}_2$ be generated by $\mathcal{G}_1$ and Φ. Then $\mathcal{G}_1 \lhd \mathcal{G}_2$, and $\mathcal{G}_1/\mathcal{G}_2$ is cyclic of order p, i.e., $\mathcal{G}_1/\mathcal{G}_2 = \{\mathcal{G}_1, \mathcal{G}_1\Phi, \ldots, \ldots \mathcal{G}_1\Phi^{p-1}\}$. So $\mathcal{G}_2 = \{\Theta^b\Phi^c \mid b < q, c < p\}$ and will satisfy the following transposition[38] conditions:

    i) $\Phi(j) = \alpha j$
   ii) $\Theta^q = I, \Phi^p = \Theta^l$ for some $l < q$
  iii) $\Phi\Theta = \Theta^x\Phi$ for some $x < q$

There exists g_2 in $\mathcal{F}_1$ such that $k^p = g_2$ for some k not in Γ_1, and $\Phi(g_2) = g_2$. So we can construct a crossed product algebra Γ_2 using the Extension Theorem as in Stage 1. This second algebra will have basis $\{j^b k^c \mid b < q, c < p\}$ over $\mathcal{F}_1$. We extended Φ from $\mathcal{F}_1$ to $\mathcal{G}_1$ then to $\mathcal{G}_2$ using the Multiplicative Extension Theorem.

---

[37]Dickson 1930, Theorem 5.

[38]By 'transposition' is meant transposing or interchanging two adjacent terms, since only the base field is abelian.

Stage 3: Construction of algebra Γ_3

Finally, let Ψ be an automorphism from $\mathcal{G}\backslash\mathcal{G}_2$, having order s. Repeating the above construction, we have a $g_3 \in \mathcal{F}_1$ and h not in Γ_2 such that $h^s = g_3$ and $\Psi(g_3) = g_3$. We extend Γ_2 to Γ_3 by adjoining h to Γ_2 using the Extension Theorem. The basis for Γ_3 will be

$$\{j^b k^c h^d \mid b < q, c < p, d < s\}$$

over $\mathcal{F}_1$. The Galois group for this new algebra will be

$$\mathcal{G} = \{\Theta^b \Phi^c \Psi^d \mid b < q, c < p, d < s\}.$$

And we extend Ψ from $\mathcal{F}_1$ to $\mathcal{G}_1$, then to $\mathcal{G}_2$, then to $\mathcal{G}_3$ using the Multiplicative Extension Theorem. The additional transposition conditions[39] needed are:

  iv) $\Phi(j) = \delta_1 j$, for some $\delta_1 \in \mathcal{F}$
  v) $\Psi(k) = \delta_2 k$, for some $\delta_2 \in \mathcal{F}$
  vi) $\Psi^s = \Theta^m \Phi^n$, for some $n < q, m < p$
  vii) $\Psi\Theta = \Theta^y \Phi^v \Psi$ for some $y < q, v < p$
  viii) $\Psi\Phi = \Theta^u \Phi^z \Psi$, for some $u < q, z < p$.

We have constructed crossed product algebras $\Gamma_1 \subset \Gamma_2 \subset \Gamma_3$ with $\mathcal{G}_1 \lhd \mathcal{G}_2 \lhd \mathcal{G}_3$ where $\mathcal{G}_1$ is generated by Θ, $\mathcal{G}_2$ is generated by Θ and Φ, and $\mathcal{G}_3$ is generated by Θ, Φ and Ψ, and with cyclic factor groups. So by taking $\Gamma = \Gamma_3$, we have our desired crossed product algebra. It will have basis

$$\{\alpha^a j^b k^c h^d \mid \alpha < qps, b < q, c < p, d < s\}$$

over the base field $\mathcal{F}$ and be of dimension $(qps)^2$.

Now we will describe the conditions needed for associativity at each stage of our construction. At each stage n, a new condition is required to relate each previous condition to the new generator. Thus in Dickson's case of a solvable group with three generators, we will need ten conditions to assure associatively. These conditions arise through the process of constructing each algebra and extending the automorphisms. More detail of this process will be given in the discussion of Rees's work.

To construct Γ_1, we require

---

[39] Recall that $\Psi\Theta$ and such represents composition.

(1) $\Theta(g_1) = g_1$

To extend Φ to $\mathcal{G}_1$, we require

(2) $\Phi(g_1) = \alpha\Theta(\alpha)\cdots\Theta^{q-1}(\alpha)g_1$

To extend Γ_1 to Γ_2, we require

(3) $\Phi(g_2) = g_2$ and
(4) $g_2 = \alpha\Phi(\alpha)\cdots\Phi^{p-1}(\alpha)\Theta(g_2)$

To extend Ψ to $\mathcal{G}_1$, we require

(5) $\Psi(g_1) = \delta_1\Theta(\delta_1)\cdots\Theta^{q-1}(\delta_1)g_1$

To extend Ψ to $\mathcal{G}_2$, we have

(6) $\Psi(g_2) = \delta_2\Psi(\delta_2)\cdots\Psi^{p-1}(\delta_2)g_2$

And to extend Ψ to $\mathcal{G}_3$, we need

(7) $\dfrac{\Phi(\delta_1)}{\delta_1} = \dfrac{\Psi(\alpha)}{\alpha}\dfrac{\Theta(\delta_2)}{\delta_2}$

Finally, to extend Γ_2 to Γ_3, we require

(8) $\Psi(g_3) = g_3$
(9) $g_3 = \delta_1\Psi(\delta_1)\cdots\Psi^{s-1}(\delta_1)\Theta(g_3)$, and
(10) $g_3 = \delta_2\Psi(\delta_2)\cdots\Psi^{s-1}(\delta_d)\Phi(g_3)$

The important result of Dickson's 1930 paper, with respect to our discussion of the development of algebras leading to Rees's work, is the new, more systematic approach to building up algebras. This was done by Dickson's explicit use of his Extension Theorem and his implicit use of the Multiplicative Extension Theorem.

By the time this paper was being published, Dickson had turned his attention (and that of his graduate students) to number theory and Waring's problem. However, when Rees arrived on Dickson's doorstep in the fall of 1929, demanding a topic in algebras, he gave her the task of verifying his method for the four generator case. No new theory was required for this, yet it was no easy task.

3.7. Rees's Doctorial Dissertation

Mina Rees's Ph.D. dissertation, "Division Algebras Associated with an Equation Whose Group has Four Generators," was submitted to the University of Chicago in December of 1931, and published by the AMS in 1932.[40] In this dissertation, she carries out the construction of an associative division algebra when the Galois group is solvable with four independent generators.[41] She assumes Dickson's construction of algebras with two and three generators, presented in his 1930 paper. She then exhibits the requirements to extend his algebra Γ_3 to the four generator case. Her work focuses on finding the additional conditions to make it an associative algebra. She concludes her work with an example of such an algebra of dimension thirty-two.

Dickson's constructions of algebras with one, two, and three generators provided a total of ten conditions for associativity. For each of these conditions, Rees will need a new condition to relate it to her new fourth generator. This gives a total of twenty associativity conditions for an algebra with four generators. These conditions come out of the extension process described below.

The construction starts as previously; a primitive element α of $\mathcal{F}$ is found that gives a normal field extension. Dickson and Rees actually use the irreducible polynomial $f(x)$ over $\mathcal{F}$ as their starting point. However $f(x)$ is found only after finding the primitive element which gives a normal field extension. Thus even though Dickson and Rees seem to start their construction by choosing an arbitrary polynomial of the desired degree, it is really not arbitrary. So to avoid confusion, we will take our starting point as the primitive element α.

---

[40][**49**].

[41]See definition of solvable in the previous section.

Now let this element have degree $qpst$. Let the Galois group $\mathcal{G}$ be such that

$$\mathcal{G} = \{\Theta^b \Phi^c \Psi^d \chi^e \mid b < q, c < p, d < s, e < t\}.$$

Assume we have the crossed product algebra Γ_3 with basis

$$\{j^b k^c h^d \mid b < q, c < p, d < s\}$$

over $\mathcal{F}$, with corresponding Galois group $\mathcal{G}_3$ generated by Θ, Φ, and Ψ as in our previous discussion.[42]

We will now construct the new algebra $\Gamma = \Gamma_4$ by adjoining automorphism χ to Γ_3, in the same manner as we did for Dickson's 1930 paper. Then Γ will be of dimension $(qpst)^2$ over $\mathcal{F}$ with basis

$$\{\alpha^a j^b k^c h^d l^e \mid a < qpst, b < q, c < p, d < s, e < t\}$$

Step 1: Extend χ from $\mathcal{F}_1$ to $\mathcal{G}_1$

Since by assumption, the group is solvable, we can use the Multiplicative Extension Theorem. Take b in the statement of the theorem to be j. Then the theorem requires

$$\chi(j^q) = (\chi(j))^q,$$

where $\chi(j) = \epsilon_1 j$, for some $\epsilon_1 \in \mathcal{F}$. Thus $\chi(g_1) = (\epsilon_1 j)^q$. Using the transposition rules (i) through (viii) from the previous construction we get the first of Rees's conditions for associativity:

$$\chi(g_1) = \epsilon_1 \Theta(\epsilon_1) \cdots \Theta^{q-1}(\epsilon_1) g_1. \tag{1}$$

Step 2: Extend χ from $\mathcal{G}_1$ to $\mathcal{G}_2$

Again using the Multiplicative Extension Theorem, we require

$$\chi(k^p) = (\chi(k))^p,$$

where for $\chi(k) = \epsilon_2 k$, for some $\epsilon_2 \in \mathcal{F}$. Then $\chi(g_2) = (\epsilon_2 k)^p$ thus

$$\chi(g_2) = \epsilon_2 \Phi(\epsilon_2) \cdots \Phi^{p-1}(\epsilon_2) g_2. \tag{2}$$

---

[42]Note that $\mathcal{F}_1 = \mathcal{F}(\alpha) = \{\alpha^a \mid a < qpst\}$ over $\mathcal{F}$.

For χ to remain an homomorphism, we need $\chi(kj) = \chi(k)\chi(j)$. Using the transposition rules we get

$$\chi(\alpha)\epsilon_1 j \epsilon_2 k = \epsilon_2 k \epsilon_1 j,$$

which gives

$$\chi(\alpha)\epsilon_1 \Theta(\epsilon_2) jk = \epsilon_2 \Phi(\epsilon_1)\alpha jk.$$

Thus

$$\frac{\Phi(\epsilon_1)}{\epsilon_1} = \frac{\chi(\alpha)}{\alpha} \frac{\Theta(\epsilon_2)}{\epsilon_2}. \tag{3}$$

Step 3: Extend χ from $\mathcal{G}_2$ to $\mathcal{G}_3$

This requires

$$\chi(h^s) = (\chi(j))^s,$$

where $\chi(h) = \epsilon_3 h$, for some $\epsilon_3 \in \mathcal{F}$. Which leads to

$$\chi(g_3) = \epsilon_3 \Psi(\epsilon_3) \cdots \Psi^{s-1}(\epsilon_3) g_3. \tag{4}$$

Similarly, $\chi(hj) = \chi(h)\chi(j)$ leads to

$$\frac{\Psi(\epsilon_1)}{\epsilon_1} = \frac{\chi(\delta_1)}{\delta_1} \frac{\Theta(\epsilon_3)}{\epsilon_3}, \tag{5}$$

and $\chi(hk) = \chi(h)\chi(k)$ gives

$$\frac{\Psi(\epsilon_2)}{\epsilon_2} = \frac{\chi(\delta_2)}{\delta_2} \frac{\Phi(\epsilon_3)}{\epsilon_3}. \tag{6}$$

Step 4: Extend Γ_3 to Γ_4

Using the Extension Theorem, we have

$$\chi(g_4) = g_4 \tag{7}$$

where $g_4 \in \mathcal{F}_1$, and $l \notin \Gamma_3$ such that $l^t = g_4$. Now we need to calculate $g_4 z = \chi^t(z) g_4$, for $z = j, k, h$. First, $g_4 j = \chi^t(j) g_4$. The transposition rules are used once again to produce

$$g_4 = \epsilon_1 \chi(\epsilon_1) \cdots \chi^{t-1}(\epsilon_1)\Theta(g_4). \tag{8}$$

$g_4 k = \chi^t(k) g_4$ gives us

$$g_4 = \epsilon_2 \chi(\epsilon_2) \cdots \chi^{t-1}(\epsilon_2)\Phi(g_4), \tag{9}$$

and $g_4 h = \chi^t(h) g_4$ provides the last condition for associativity,

$$g_4 = \epsilon_3 \chi(\epsilon_3) \cdots \chi^{t-1}(\epsilon_3) \Psi(g_4), \tag{10}$$

Thus these ten new conditions for associativity, (1) through (10), of Rees's combine with the ten previous conditions found by Dickson, to give the twenty conditions needed for Rees's algebra Γ to be a crossed product.[43]

By way of summary, the twenty conditions needed to make Γ a crossed product, by inspection, fall into three types of categories:

I. $\Theta(g_1) = g_1$

II. $\dfrac{\Phi(\delta_1)}{\delta_1} = \dfrac{\Psi(\alpha)}{\alpha} \dfrac{\Theta(\delta_2)}{\epsilon_2}$

III. $\Phi(g_1) = \alpha\Theta(\alpha) \cdots \Theta^{q-1}(\alpha) g_1$ or $g_3 = \delta_1 \Psi(\delta_1) \cdots \Psi^{s-1}(\delta_1)\Theta(g_3)$.

Using our transposition conditions, the equations in each category can be seen to be notational variants of each other. So in essence, we have only three conditions which are necessary for Γ to be a crossed product. However Dickson does not state this. Dickson's Theorem 4 of his 1930 paper shows that these conditions are not only necessary, they are also sufficient for the constructed algebra Γ to be associative.

Rees concludes her dissertation with a simple concrete example. By taking $q = 4$ and $p = s = t = 2$, she gets an algebra of dimension thirty-two. The

---

[43]Note that the order in which the automorphisms are adjoined is arbitrary. Adjoining them in a different order will also result in a (possibly different) crossed product algebra.

transposition conditions then reduce to:

$$\Theta^4 = \Phi^2 = \Psi^2 = \chi^2 = I$$
$$\Theta\Phi = \Phi\Theta$$
$$\Theta\Psi = \Psi\Phi\Theta$$
$$\Phi\Psi = \Psi\Phi$$
$$\Theta\chi = \chi\Phi\Theta^3$$
$$\Phi\chi = \chi\Phi$$
$$\Psi\chi = \chi\Psi$$

All the other parameters are taken to be 0 or 1.

Rees, like Dickson, speaks of her algebras as if they are division algebras. Note, however, that her example above is not a division algebra. Let $\mathcal{F}$ be the base field, and let $k^2 = h^2 = 1$ in $\mathcal{F}$. Then

$$(k+h)(k-h) = k^2 + kh - kh - h^2 = 1 + 0 - 0 - 1 = 0.$$

Since k and h are basis elements, $k+h$ and $k-h$ are linearly independent. Thus $k+h$ and $k-h$ do not equal zero. So we have non-zero divisors of zero, and the algebra is not a division algebra.[44] It appears that after the 1926 paper, Dickson, and then Rees, assumed divisibility.

Rees's work is very clear and well organized. She does a beautiful job of pulling Dickson's work together into a general and a specific example. At the start of her paper she states her assumptions, which are primarily Dickson's 1930 construction. She uses the Extension Theorem explicitly, but her use of the Multiplicative Extension Theorem is implicit, and she does not explain the extension process used. The addition of this would have made her paper a very nice capstone to this line of research indeed.

≪   ≫

As stated earlier, Dickson had for the most part concluded his work on division algebras by the time Rees arrived in Chicago in 1929. His 1930 paper may already have been at the printer. Dickson's introduction to his 1930 paper indicates that he had postponed, possibly indefinitely, the case of four generators.

---

[44]This was noticed by William Howard.

When Rees arrived asking for a topic in division algebras, Dickson may have seen this as a dissertation topic beneficial for both of them.[45]

At this time, as Saunders Mac Lane described,[46] new Ph.D.s were placed in positions via the old boy's network. Students were often viewed in one of two ways – either as future researchers or future educators – and their training and research topics were picked accordingly. Since Rees had a good teaching position waiting for her at a small women's college, Dickson may have placed her in the later category. Her dissertation is very well executed, precise, and clearer than Dickson's work. However it does not appear to be a good topic to launch a person's research career. Her dissertation seems to be more of a task Dickson set for Rees to complete. It gave Rees a dissertation topic, so she could get back to Hunter.

For Dickson, having Rees write up the details of the four generator case provided nice closure to his research in associative algebras. It verified that the method could be extended one more step, and hence, presumably, any number of steps. Our discussion of Rees's dissertation does not show the sixty-one different calculations she made and equations she deduced in order to achieve her results. Thus without a more abstract method (developed by Noether at about this time), or computers, the case of five or more generators would be little more than an exercise in frustration.

Dickson's concrete methods made further progress in the theory unmanageable. The abstract methods of Noether's group became the standard approach in the field. Saunders Mac Lane commented that, "[while in Göttingen, 1931] I listened to Noether's ... lectures on hypercomplex systems. I was startled to find that she often cited an influential German translation, *Algebren und Ihre Zahlentheorie*, of Dickson's book.[47] ...I heard from Noether about the use of factor sets, ... they made Mina's thesis obsolete."[48] In fact, it seems unlikely that anyone has used Rees's work. Mac Lane noted that he was surprised when, in arriving in Chicago in 1930, Noether's work did not seem to be known.[49] This

---

[45]The author surmises that this may be why Rees was not given a preliminary trial problem to complete before Dickson took her on as a student.

[46][**34**] and [**29**].

[47][**12**]

[48][**29**, p. 870].

[49]*Ibid.*, p. 870.

is curious since Dickson was in contact with European mathematicians through-out his career. So it looks as if information was flowing from the Americans to the Europeans, but not the other way in this instance. Historians Judy Green and Jeanne LaDuke report that in 1933 Rees obtained Oskar Perron's *Algebra* that presented the Noetherian approach. After reading this, Rees realized her knowledge of algebras was not sufficient to pursue more research in that field.[50] So she may have decided at that point to put her energies towards teaching and ultimately administration.

In Rees's memorial, Dennis Sullivan commented, "Mina's research was somewhat curtailed because she was preempted by the great work of Brauer, but it did show that she had very good taste: she chose this problem herself and even convinced Dickson to work on it again."[51] After receiving her doctor-ate, Rees returned to Hunter College in New York as Assistant Professor. She continued at Hunter until the United States joined the Second World War.

---

[50][**30**].
[51][**26**, p. 17].

CHAPTER 4

World War II

"The decisive event in my life came in 1943, when I accepted a
wartime job that introduced a whole new orientation into my
career."
  –Mina Rees, 1979[1]

In 1943 The Applied Mathematics Panel was established under the National Defense Research Committee (NDRC). The panel was headed by Warren Weaver, who immediately chose Rees as one of his technical aides, and soon after, his executive assistant. At the time of the Panel's disbanding at the end of the war, Rees had the 'simulated' rank of Admiral in the Navy. In these capacities, Rees learned the workings of the American research community, and the keys to effective administration. In order to understand Rees's role in war-time research, an understanding of the organization and workings of the AMP are needed.

4.1. The Applied Mathematics Panel

Following closely on the heels of Pearl Harbor, the U.S. Office of Scientific Research and Development was reorganized in 1942. Warren Weaver described the need for the reorganization as follows:

> As the war went on, the emphasis [by NDRC] on the design
> and production of hardware necessarily tapered off somewhat,
> for the practical reason that by then a brand-new device simply
> could not be conceived of, designed, built in pilot model, tested,

---

[1][**77**, p. 15].

improved, standardized, and put into service in time to affect
the conduct of the war.[2]

As part of this reorganization, the NDRC, which had been established in 1940,
was also reorganized. This was done prior to the U.S. entering World War II, to
lend scientific support to the military, in the event the U.S. entered the war. The
NDRC was comprised of groups of scientists and engineers working on issues
related to such things as submarine warfare, radar, electronic countermeasures,
rocketry, and explosives. The next year, Vannevar Bush, OSRD's new director,
established the Applied Mathematics Panel. Warren Weaver, head of a section
of the Fire Control Division of NDRC, became the chief of the AMP. Weaver
(1894–1978) had been chair of the mathematics department at the University
of Wisconsin Madison before joining the Rockefeller Foundation in 1932 as
Director of it's Natural Sciences Division. He continued in this position until
1959.

The AMP was a panel, and not a division of the OSRD. The distinction
being that a panel is a group of people available to work on many different problems as they arose, where as a division had a set assignment. The Panel was
made up of government appointees such as Thornton Fry, Marston Morse, Griffith Evans, Howard Robertson, Abraham Taub, Oswald Veblen, and Richard
Courant and was headquartered in the Empire State Building in New York
City.[3]

The Applied Mathematics Panel was established to provide mathematical
support in solving the problems faced by the military and other OSRD divisions
during the war (there had been no mathematics group in the OSRD). In particular, the AMP would work on problems of air warfare and the improvement of
theoretical accuracy of equipment. Rees described that the Panel's task "was to
help with the increasingly complex mathematical problems that were assuming
importance and with those other problems that were relatively simple mathematically but needed mathematicians to formulate them adequately."[4] The
OSRD did have mathematicians on the staff, but they were few in number and
too dispersed geographically to attack the variety of questions beginning to be

---

[2][**82**, p. 276].
[3]*Ibid.*, p. 275.
[4]*Ibid.*, p. 277.

asked as the U.S. involvement in the war became eminent. The AMP represented the largest group of mathematicians working for the war effort; providing consultants having wide mathematical backgrounds. Most of these mathematicians were academics conducting pure research who volunteered or were asked to join the war effort. They came from many mathematical disciplines, and worked to formulate solutions as quickly as possible to problems that arose during the war.[5] Rees stated that by 1946, when the AMP was disbanded, over 200 studies had been undertaken by the Panel. Nearly half were the result of requests by the military.

The focus was on getting results usable in the war effort. Thus the AMP was not a research program as such. Education and research for the sake of research were not part of the AMP goals. That would come later as the work of the AMP rolled over into the Office of Naval Research after the war, and eventually with the establishment of the National Science Foundation in 1950. The AMP did, however, provide specialized technical training for small groups of military personnel in order to facilitate the implementation of new techniques and theories developed by the AMP.

One of the primary jobs of the Panel was to write and oversee contracts between the government and various universities. The OSRD used the method of procurement contracts that the government used to fund research and to purchase goods and equipment. Thus the research was conducted under government contracts and not grants, fellowships, and the like. The Panel was given problems with which the military needed help. It then determined the possibility of solution " ... often [we] were asked to perform a miracle which wasn't quite within our capacity" stated Rees.[6] If the Panel felt a solution was possible, it then determined the mathematical nature of the problem. Since problems from different divisions of the OSRD or the military may have the same or similar mathematical underpinnings, the AMP was also a communications link between various divisions. Rees referred to the AMP as a "clearing house for information." The problem was then passed on to one of several university research groups or the Mathematical Tables Project.

---

[5]At that time there was not the distinction there is today between pure and applied mathematics.

[6][**39**].

To get a better feel for the work done note that the *Summary Technical Report of the Applied Mathematics Panel* of 1946 listed the following categories of research conducted: aeroballistics, theory of deflection shooting, pursuit curve theory, design and characteristics of own-speed sights, lead computing sights, basic theory of a central fire control system, analytic aspects of testing airborne fire control equipment, and new developments such as radar and stabilization.[7]

The research groups were located at eleven institutions: Princeton, Columbia, New York University, University of California Berkeley, the Franklin Institute, Brown University, the Institute for Advanced Study, Harvard, the Carnegie Institution of Washington, the University of New Mexico, and Northwestern University.

The university groups were organized around different concepts. The NYU group, headed by Courant, worked on problems of gas dynamics, for use in both air and water explosions, and jet and rocketry theory. Out of this work came the Shock Wave Manual in 1944, which gave a mathematical theory of fluid compression. Research at Brown, headed by Richardson, centered on classical dynamics and deformable material. The Harvard group worked on underwater ballistics. Columbia was the largest contract university, with three groups, the Statistical Research Group, the Bombing Research group, and the Applied Mathematics Group. The Applied Mathematics Group at Columbia worked on such topics as air ballistics, deflection, curve theory, and central fire control. Rees's University of Chicago classmate Saunders Mac Lane was the Technical Representative of this group.

Besides obviously war-related pursuits such as effective equipment use and fire control, operations research became a field in its own right because of the war effort. In 1944, Warren Weaver wrote,

> It has become quite clear that the emphasis ... should, from now on, shift rather rapidly from long range fundamental studies (of admitted importance but dubious application to this war) and even from analysis studies of specific mechanisms and devices ... to activities of the 'operational research' type,

---

[7][82].

related as closely as possible with the practical day-to-day conduct of the war.[8]

The Air Forces[9] gave a large impetus to this branch of research. It was concerned with insuring the availability of material, men, and resources in a timely
manner. This lead to research in linear programming carried out at the Mathematical Tables Project under Arnold Lowan, using the new methods of George
Danzig. Danzig's simplex method was developed and tested there. Thus the
study of optimization of supply and demand was an AMP project. The Air
Forces also wanted operations research applied to air warfare. The groups at
Columbia and Northwestern worked on this. The Applied Mathematics Group
at Columbia had a special program to train mathematicians to do operations
research overseas. In particular, the 8th Air Force was in England, where American and English operations research specialists worked together to develop the
field.

Another important outcome of the research contracted by the Applied
Mathematics Panel was statistical research. The statistical groups were lead by
Jerzy Neyman at the University of California at Berkeley, Sam Wilks at Princeton, and Allen Wallis at Columbia. Of concern to these groups were problems
such as bombing, industrial techniques, damage probability studies, statistical
methods of inspection and sampling. This arose from a concern on the part of
the Navy about the destructive sampling of munitions, and led to the study of
sequential analysis.

Arguably the most important research initiative on the part of the AMP was
early work on computers, and computing. Work on analog computers primarily
grew out of the Air Forces' need to improve fire control (digital computers were
developed later when faster calculations were needed during work on the atomic
bomb). Fire control deals with the ability to aim accurately from a moving
aircraft, and the development of electrical anti-aircraft directors. The Air Forces
had simulators that they used for training and to increase the effectiveness of fire
control, but these simulators were not adequate. Thus the need to devise better
simulators led the AMP to look into better and better electronic programs for

---

[8][**83**, p. 191].

[9]The Air Force would not be established until after the war, so Air Forces refers to the
aviation branches of the existing military branches.

the simulators, which eventually lead to work on the first analog computers. On the other hand, work was being done on the first electronic computer, the ENIAC,[10] at the University of Pennsylvania, with John von Neumann.

The Applied Mathematics Panel also oversaw the Mathematical Tables Project. This project came into being in 1938 as a result of the Works Progress Administration and was originally administered by the National Bureau of Standards. Most of the hand-computing needed during the war was conducted by the Mathematical Tables Project, work that would be taken over by the new digital computers developed immediately after the war. Much of the work in operations research was conducted here also, as described above.

Since the majority of the research was done at universities, the viewpoint of academia gave new insight into war related research. Many times the academic researchers were able to view a problem in a different light than the primarily engineering focused military researchers. They also looked at the problem in different ways from each other. Rees recalled,

> [O]ne of the things I found interesting was the way Adrian Albert, for example, would solve a problem in contra-distinction to the way Saunders Mac Lane would solve the same problem ...I think this was one of the most interesting things mathematically: that the same problem would be attacked typically, not surprisingly, but typically, by the resources that [a] particular mathematician had most completely under control, and often solved by two different machineries.[11]

The academic setting also inspired research into the theory as well as the solution of the problem.

The university groups were chosen on the basis of the reputation of the mathematicians, and their desire to join the war effort. Most of them were not applied mathematicians. Thus the AMP provided much impetus into bringing applied and pure mathematics together. Most of its mathematicians were of the pure sort, who wanted to contribute to the war effort. Rees described them

---

[10][**40**]. In 1949, the ENIAC computed π to 2,037 decimal places!
[11][**39**].

as pure mathematicians whose results had applications. Larry Owens stated, "It [AMP] accelerated the institutionalization of applied mathematics through its support of groups like Courant at NYU and Richardson at Brown."[12]

4.2. Rees's Work on the Applied Mathematics Panel

In 1943 Rees took her second leave of absence from Hunter College to become a technical aide to the Applied Mathematics Panel, and executive assistant to Warren Weaver. Rees grew to know Weaver well during their time at the AMP. In 1987 Rees wrote the article on Warren Weaver for the series *Biographical Memoirs*.[13] It is an informative and articulate article which clearly portrays her regard, admiration, and warmth for him. In this article, she mentions how she came to be suggested for the position of technical aide.

> [I]t was Courant who was responsible for my being invited to join the staff of AMP as a civil servant. It was not because I was an applied mathematician ... it was not because I was a woman – there was no equal opportunity then. It was the good old buddy system.

Along with knowing Rees through the New York mathematical meetings, Courant had a graduate student (Bella Manel) who had taken all but one of her undergraduate mathematics courses with Rees at Hunter. Courant also knew that not only did Weaver need technical aides with mathematical backgrounds, they had to be good administrators and communicators. Courant knew that Rees was very strong in all these areas.

Rees was not looking to change careers. However, for months before the AMP offer came, Rees was becoming more and more interested and even anxious to help in some way with the war effort. When the offer came, she felt it was an opportunity not to be missed.

> I suffered for six months [before joining the Panel] trying to figure out how could I do something to contribute something

---

[12][**46**, p. 300].
[13][**81**].

to the war effort. I'm sure that most of these people went
through this too. It's a dreadfully frustrating sensation [when]
you think you ought to be able to do something, and you don't
know how to get in there and do it.[14]

Rees's job had many facets. As Weaver's executive assistant, she was the
secretary to the Panel. She attended all the Panel meetings " ... which [being
secretary] gave me a central opportunity to understand about all the problems
that came in either to Warren Weaver or to me."[15] At one point, Weaver was
incapacitated by an ear infection, and Rees filled in for him on a day-to-day
basis. After this she emerged as the representative of both Weaver and the
Panel to both the contract universities and the military.[16]

As a technical aide, she was a liaison between the research groups or contractors, and the Panel. In this capacity, she represented the government in its
dealings with the contractors. Rees believed this was the origin of university
contractors.[17] She visited military sites to gather information about what was
needed and to clarify the problems being researched. She visited the research
groups to make sure the work was moving ahead and to make sure the work
being done was useful. She helped determine whether work could be done on
the problems, and made weekly recommendations on this to the Panel. Finally,
she helped write, publish and circulate short term and long term reports. Rees
recalled that there were, "endless administrative details to be handled as the
Panel carried on a broadly based effort to bring mathematics to the service
of the war effort. When the time came to close up shop, the Panel had some
spectacular successes. Its work was much appreciated."[18]

As might be expected, Rees did experience some negative bias due to her
gender. The first being at time overt resistance to her role and authority, both
when she was suggested for the position and after joining the Panel. The male
dominated worlds of mathematics and science, Washington bureaucracy, and
the military were unaccustomed to women in high ranking positions. Especially
not women with Ph.D.s in mathematics. She also became frustrated by the

---

[14][**39**].
[15]*Ibid.*
[16][**84**].
[17][**39**].
[18][**77**].

unintentional slights she received due to the traditional roles women took. In particular she was regularly mistaken for Weavers secretary, as opposed to an actual member of the Panel. The following candid correspondence between Rees and Richard Courant is a rare glimpse into not only Rees's personal experiences as a member of the Panel, but also their friendship and candor.[19]

≪   ≫

12/19/43

To: RC
From: MR

I want to tell you about something which is making life a bit difficult for me, and to ask a very, very slight favor of you.

As you know, I am accustomed to the lordly status of a college professor. Moreover, at Hunter my position is unassailable, I expect to participate in important decisions, and no one would dream of asking me if I were somebody's secretary.

I grow a bit weary of that assumption in my present job (and of the question, "are you Weaver's secretary?"), the more because it's so near the truth. My only consolation is that, if I didn't do a lot of the things I do, WW would; and saving his time and energy doubtless justifies using mine. But the loss of prestige and professional dignity, and of the opportunity to use what special knowledge and ability I have seems a high price to pay for a sense of doing something useful toward winning the war, and for the very real pleasure I get out of many aspects of the work. This is not a complaint, – I wouldn't give up this job for the world - only a request.

For you, the moral is this: Will you please, in the future, when you introduce anyone of importance to me, say in a loud and clear voice, "This is Dr. Rees."? Your muttered "This is our boss" has consistently brought forth the assumption I speak of, and I can't take it.

---

[19]Rees/Courant correspondence courtesy of the NYU Archives

I want to be sure you understand that I find working with
WW a real privilege. He is incredibly generous and considerate.
None of my troubles are in any sense attributable to him, but
just to my being a woman.

≪   ≫

December 22nd, 1943

Dear Mina:

Only after we parted last Monday I really read and under-
stood your letter. My first and last reaction, I must admit, is
that I am glad and grateful for your frankness. Between less
good friends a situation might have developed where nothing
is said and an unexplainable state of irritation and uneasiness
slowly develops. I hope this will never happen between us even
though I can not promise never to do or to say anything tactless
again.

On the other hand, I think that you are really too sensitive
and that there is nothing farther from the thinking of anybody
in contact with you than the idea of relegating you into a place
of secondary importance or responsibility. You know just as
well as I how much your work for the Panel is appreciated.
Weaver more than once has expressed his horror at the idea
what would have happened if you had not come. I am certain
that you will feel less and less cause for your understandable
irritation but I can only promise that I shall try my best to
contribute to a happy development.

With best wishes,

Your, [Richard Courant]

≪   ≫

Even with all the successes of the AMP, and her growing recognition of the
important role she played on the Panel, she would continue to experience gender
bias when she transitioned over to the Office of Navel Research. Though the
benefits she gained while on the Panel far outweighed the personal frustrations
caused by any discrimination.

Rees describes the knowledge she gained while working with the AMP:

> [Working on the AMP] was decisive ... because it greatly
> broadened my awareness of unfamiliar fields of mathemat-
> ics and my contacts with mathematicians; and, ... because it
> greatly increased my understanding of the character and ac-
> tivities of many of our major educational institutions and of
> the structure and operations of the government including the
> military establishment ... It gave me familiarity with the work
> of many of America's most able mathematicians; and it gave
> me considerable understanding of the changes that were occur-
> ring in mathematics as a result of experiences in World War
> II ... In short, it gave me the kind of experience that made it
> appropriate for me to be invited to become head of the mathe-
> matics research program of the Office of Naval Research when
> that Office was established after the war.[20]

A 1970 Article in *Science* nicely describes the influence of her war work on
Rees's career.

> Now deeply involved in the successes and failures of mathe-
> maticians trying to make useful contributions under wartime
> exigencies to the problems against which they had been pitted,
> individually or in organized teams, she acquired an unparal-
> leled perspective on the fundamental character of mathemat-
> ics on the sources of its power, the manner of its workings,
> and the basis for the need of its use in human endeavors of
> contemporary sophistication.[21]

Rees went on to delineate the changes in areas of research induced by research conducted during the war: "the emergence of mathematical statistics
in its great variety; the development of the computer and the need for exten-
sive work in mathematics to insure a sound exploitation of the potential of
the computer; the clear opportunity to extend the use of operations research
to important new areas; the potential, through the use of computers, for new

---

[20][**77**, p. 15].
[21][**91**, p. 1149].

applicable results in analysis." Rees continued to assist the war effort through
the Applied Mathematics Panel until its disbanding in 1946.

≪   ≫

World War II had several effects on scientific and mathematical research in
the United States. It brought pure and applied scientists together, it brought
government involvement and funding into academia, and it influenced the research that was done and how it was conducted. In particular, war related
work showed that many problems need cross-disciplinary approaches to achieve
viable solutions. These changes in approach would be continued and further
developed directly after the war through such resultant agencies as the Office
of Naval Research. Rees explained the change in research brought on by the
war in an interview with Uta Merzbach. "The real difficulty that we suffered,
I think, before World War II, arose from the nature of the problems to which
mathematics was applied. These needed analysis. Now the problems need a
great variety of mathematical techniques."[22]

Many of the lines of research started or enhanced by the Applied Mathematics Panel continued after the war. Many universities set up programs or even
separate departments in these new fields, such as operations research, computing, and the applications to business of certain theories such as modeling. The
new interest in applied mathematics had a profound effect on mathematics and
mathematics departments. Areas of study such as logistics as well as the fields
specifically mentioned above found support from agencies such as the National
Bureau of Standards and the newly emerging research divisions of the armed
forces. The most influential of these was the Office of Naval Research.

Rees was an integral component of the Applied Mathematics Panel. She
was an effective bridge between research and academia on the one hand, and
the government and military on the other. She felt she had a unique position,
in that being a relatively high-ranking civilian in the military establishment,
and a woman, she could ask harder questions and push harder for what she felt
was needed. Her experience at the AMP taught her how to balance the needs
and concerns of various parties, and how to arrive at solid working solutions.

---

[22][39].

She also gained a broad knowledge of mathematics, mathematicians, and mathematical research: knowledge she would need to be an effective administrator and proponent of research. Most people who worked with her felt that though she was sometimes hard and determined, she was fair and always had the best interests of science in mind.

Due to the war and Rees's efforts to engage mathematicians in problems of immediate importance, "...mathematics had been recognized, once and for all, to be important – like physics, chemistry, and engineering mechanics – and was so treated, albeit at the price of no longer being allowed to confine itself to being pure." As Rees prepared to return to academia, she would continue to promote this 'new' field of applied mathematics. Although pure mathematics had been her love in college and graduate school, while working in the war effort she found the use of science to affect national issues to be a worthwhile and engaging occupation. She continued to promote pure research throughout her career, but her personal interests had turned to applications.

In 1948, both Warren Weaver and Mina Rees received the King's Medal of Great Britain and the President's Certificate from Harry S. Truman, for their service to the war effort.[23] After the war, Rees wrote several articles of a historical nature about the research done during and immediately following the war.[24] She was concerned that the advances made during the war be recorded accurately. She was also a member of the Mathematical Association of America's committee on World War II history. With the end of the war came the end of the Applied Mathematics Panel. But Rees's expertise was too valuable, and she was asked to join the Office of Naval Research within weeks of its conception.

---

[23]See Appendices C and D for copies of these awards.
[24][**53**], [**75**], [**77**],[**78**], [**79**], [**80**], and [**81**].

FIGURE 4.1. Richard Courant and Mina Rees at Courant's retirement party from NYU, June 1965. Courtesy of the Archives of the Graduate Center of CUNY.

FIGURE 4.2. Richard Courant and Mina Rees at Courant's retirement party from NYU, June 1965. Courtesy of the Archives of the Graduate Center of CUNY.

CHAPTER 5

Post-War: 1946–1953

"The wartime performance of these [AMP] scientists won for science the high regard of the military establishment and of Congress and the recognition that post-war expansion of research in the sciences was a national requirement." observed Rees.[1] The U.S. government saw this, and in order to keep the interaction of science, mathematics and government alive, launched a program of federal support of research never seen before. The lead in this endeavor was taken early by the Navy. "It was clearly seen by the government and those responsible for the armed services that a large scale fostering by the U.S. government of fundamental research, the basis of all research, was unavoidable. Only thus could we hope to hold our own in years to come, and incidentally build up a suitable reserve of talented men for emergencies. This was actually acted upon by the Navy who thus took the lead by some years with the creation of the Office of Naval Research. Needless to say, as the purest of all sciences, mathematical research might well have lagged behind in such an undertaking."[2]

Rees played an invaluable role in setting the new policies that guided research in the mathematical sciences in the crucial post war years. However, pure mathematics could easily have been left behind in light of the more obvious accomplishments of 'applied' mathematics during the war. In fact, the American Mathematical Society resolution goes on to state, "That nothing of the sort happened is beyond any doubt traceable to one person—Mina Rees." 'Fundamental research' became the focus, as opposed to the development and application of known results, as primarily done during the war. This interaction of government and academia, started during the war, changed the lives of

---

[1][**75**, p. 102].

[2]Resolution of the Council of the AMS, 1953. See Appendix F for the full resolution.

scientists, the research they conducted and how it was conducted. A new system developed which would become the basis of our current system of federal support for research.

Although she is most widely known for her work in early computing during this time, during the post-war period Rees had wide-reaching influence on mathematical and scientific research through her work as head of the Mathematical Sciences Division and later as the Deputy Science Director of the Office of Naval Research.

5.1. The Office Of Naval Research

When the Office of Scientific Research and Development was established in 1946 to aid the war effort, it was with the understanding that it would be terminated with the close of the war.[3] As the end of hostilities neared, government leaders were concerned about losing the momentum that research had attained during the war. They needed to encourage researchers to continue to contribute to the solution of problems of national interest. This concern was voiced by the Secretary of the Navy, James V. Forrestal, in his annual report to President Truman in 1946.

> In peace, even more than in war, scientists owe to their nation an obligation to contribute to its security by carrying on research in military fields. The problem which began to emerge during the 1944 fiscal year is how to establish channels through which scientists can discharge this obligation in peace as successfully as they have during the war ... The Navy believes the solution for the problem is the establishment by law of an independent agency devoted to long-term, basic, military research, securing its own funds from Congress and responsive to, but not dominated by, the Army and Navy ... The Navy so firmly believes in the importance of this solution to the future welfare of the country that advocacy of it will become settled Navy policy ... The Navy feels so deeply about the importance

---

[3]Much of the material in this section is from Rees's 1977 article, "Mathematics and the Government: The Post-war Years as Augury of the Future" [**75**].

of the solution of this problem that it requests your intervention, guidance and support on this problem, which transcends the responsibility and authority of any single department.[4]

Rees described the government's view on the American research community as follows.

> At the end of World War II the urgent need of the Government for new scientific results, sometimes in narrowly delimited fields, and the recognition that the universities of the United States were not planning and were financially unable to embark on a program of basic research in the sciences adequate to provide for these needs, led to the decision to initiate a widespread program of Federal support to university research in the sciences.[5]

She goes on to state two reasons for this support; the need for expensive equipment, and the need for results in particular areas of research.

Many government leaders, scientists and mathematicians were consulted about the development of federal funding of research by the military. One of these was Vannevar Bush, whose 1945 book, *Science, the Endless Frontier,* helped guide the establishment of the National Science Foundation in 1950. Another person approached for input was Mina Rees. Rees was asked about her views on establishing an office within the Navy to fund university research in mathematics. "I expressed grave doubts. I thought it unlikely that mathematicians would be enthusiastic about receiving money from the government to support their peacetime research and even more unlikely that money from one of the military services would be welcome."[6]

Overall, support within the military for the new agencies was strong. The Office of Naval Research was established by an act of Congress on August 1, 1946, by Public Law 588. This law stated in part that the purpose of the ONR was, "to plan, foster, and encourage scientific research in recognition of

---

[4][**75**, p. 104].

[5][**63**].

[6][**71**, p. 179]. This is still a concern within the government and academia today.

its paramount importance as related to the maintenance of future naval power, and the preservation of national security ... "[7] Government funding of basic research at U.S. universities had started with the Navy in the late 1930s. The Navy stated that, "by 1945 the Navy was the major U.S. government agency funding basic and fundamental science."[43] The Office of Ordnance Research of the Army and the Office of Scientific Research of the Air Force were established soon after the ONR to play the same role as the ONR did for the Navy.

Rees was immediately asked to join the ONR[8] as head of its Mathematics Branch[9] in Washington D.C. After consulting friends and colleagues, particularly Warren Weaver and Richard Courant, and though still leery, she decided to participate in this 'uncertain venture'. Rees was the first woman in this new and growing field of grant administration that developed after the war. She would advance to become the Director of the Mathematical Sciences Division in 1949, and finally Deputy Science Director before leaving the ONR to return to academia in 1953.

The following story of Rees's move from New York to Washington D.C. to join the ONR is one of her favorites.

> When I arrived in Washington in August 1946, it was impossible to find a place to live. No apartments were available and most hotels permitted a guest to stay only five days. When I found one [The Lafayette] that extended its hospitality for two weeks, I was enchanted. I made virtue of necessity and, every two weeks, vacated my room and went on a trip to a leading mathematics department. On my return I registered for another two weeks.[10]

Through these trips, Rees was able to talk with many of the country's prominent mathematicians about the state of research and their views on funding. She found that scientists and researchers were supportive overall, however,

---

[7][36, p. 45].

[8]Rees comments in her 1979 AWM letter that there was some concern over her nomination due to her gender.

[9]The other ONR branches were the Computer, Logistics, and Mechanics Branches.

[10][75, p. 105].

university presidents were unsure about writing Naval support into their budgets. Along with determining the level of interest in Naval funding, she hoped to answer the following question; "why did you [the federal government] give financial support to mathematics?"[11] The answer she got was that mathematicians needed time to do mathematics, and that more mathematicians were needed. Using these ideas as a guide, Rees developed the mathematics program of the ONR. The information she gained from her trips would also be invaluable to her during her tenure at the ONR, as she tirelessly strove to balance government need and policy with support of mathematics and the mathematical research community.

Weyl reported that, "ONR made her [Rees] the architect of the first large-scale, comprehensively planned program of support for mathematical research; she pioneered its style, scale, and scope."[12] In her new role, Rees made funding decisions and set policy. She also made sure that projects moved forward and that important scientific studies were undertaken and published. She traveled widely across the country to consult with researchers and monitor progress. These travels also served to promote the ONR and encouraged researchers to become involved with ONR contracts (and to reassure them "of the pure intentions of the ... ONR"[13]). She traveled to Europe to oversee joint projects between the U.S. government and other countries. All this allowed her to keep up with research being conducted at home and abroad. Rees and her office coordinated various conferences and seminars related to the various areas of research supported by the Mathematics Branch. She was also a member of several advisory committees on behalf of the ONR.

Fairly early on, it became clear that support of pure mathematical research, as opposed to research with strictly military applications, would also fulfill the government's goal of creating a strong country through strong research. This change in policy allowed Rees to fund support of pure research through the Mathematics Branch. Rees described a crucial moment in her efforts:

> One night early in my tenure [at ONR] I was sitting at my desk, working late, when I was joined by the military officer whom

---

[11][**90**].
[12][**91**, p. 1149].
[13][**83**, p. 216].

the staff of the research division identified as the spiritual father
of the Office of Naval Research, Capt. Robert Conrad. He was
a great man and a great leader, and his energy and enthusiasm
set the tone of ONR. He sat down, and said to me, after a little
chit-chat: "Mina, if you want to include pure mathematics in
your program, I'll support you in your decision." This was a
great day for all of us, for it meant an end to the constant worry
as to whether the Navy would see the needs of mathematics as
we saw them.[14]

Frederick Terman, the Dean of the Engineering School at Stanford during
this time, summarized the situation as follows: "[I]n the critical half dozen years
immediately after the end of World War II the Office of Naval Research was
virtually the only source of funds available for the support of basic research in
the mathematical sciences."[15]

Of interest is Rees's impact on research in the United States: not just in
what research was conducted, but on how it was and still is conducted. This
impact was felt primarily in the form of support. The Office of Naval Research
was not authorized to support educational programs. Thus it could not provide funds in the form of fellowships directly to researchers. Rees supported
research by providing support through the following, in many cases previously
unknown devices: research assistantships for graduate students, research asso-
ciateships for post doctorates, secretarial assistance, travel support, summer
salaries, sabbatical leaves, release time, and funding for the journals *Mathematical Reviews* and *Applied Mechanics Reviews*. She believed that support from
mission-oriented agencies was more advantageous than a diversity of funding or
larger overall amounts. "... [There are] advantages of relating some research to
questions that must be answered and areas of study in which more knowledge
is needed if we are to advance our social purposes."[16]

Rees also believed that in order to conduct good research, one needed the
freedom to experiment and make mistakes. She incorporated this view into her
funding, allowing her researchers room to explore by giving them a generous

---

[14][**75**, p. 107].
[15]*Ibid.,* p. 106.
[16][**71**, p. 22].

amount of control in the areas of time and focus of research. At the same time she expected high quality work from herself, her staff, and her researchers. One biographical article about Rees says the following: "One of the outstanding achievements of Mina Spiegel Rees is that she created an environment in which great things could be accomplished, both by herself and by others."[17]

It is hard to think of academic life today without the forms of support listed above. Much of the support money went to young mathematicians. The ONR felt that new and promising researchers needed to be encouraged to continue in research and contribute to the scientific strength of the United States.

Mary Cartwright is an example of how ONR support benefited young researchers. At the invitation of Rees, Cartwright came to the United States from England in 1949 to work in nonlinear differential equations. Her ONR contract allowed her to spend time at Stanford, UCLA, and Princeton.[18]

In order to select new Ph.D.s to receive ONR funding in the form of post-doctorial research positions, the ONR went to the Mathematical Advisory Committee of the National Academy of Sciences for recommendations. The committee, consisting of John von Neumann, Griffith Evans, Harold Morse, Marshall Stone, Hassler Whitney, and Oscar Zariski, used peer review to choose young mathematicians to receive one-year contracts and to decide which researchers or research groups to support. In every case, the proposals originated with the researcher. Rees felt that the system could not have worked without the generous support of the mathematical community, especially in the way of reviews. Since Rees knew most of the prominent mathematicians in the country, she felt that getting peer review was not difficult. There were no geographical restrictions on granting contracts. However the ONR did try to fund mathematicians

---

[17][**45**, p. 174].

[18][**38**]. The author's doctorial dissertation advisor, William Howard, was also the recipient of funding from the ONR at this time. While a graduate student at the University of Chicago, he received a generous ONR stipend for the academic year 1952–53 to be Andre Weil's research assistant. (Weil told him that his duties consisted of making sure that the current literature seminar was supplied with chalk. The point being that Howard's duties were to attend the seminar and absorb as much knowledge as possible.) For the academic year 1953–54, he received a post-doctoral contract from the ONR to study differential equations and operator theory at Columbia University and to attend Richard Courant's seminars at NYU. The contract was renewed for a second year, during which he returned to the University of Chicago to participate in Marshall Stone's differential equations seminar.

in areas of the country that did not have strong research departments, in the hope of creating new centers of research. And the impact of this policy can be seen in the strong research programs at institutions in all regions of the United States, as opposed to primarily the two coasts prior to the war.

Another of the many influences of this new policy was on the organization of university departments. The method of funding gave researchers, instead of the administration which had traditionally made funding decisions, the power to set research and educational priorities. Rees stated that within a few years of the development of the ONR, working conditions, salaries of researchers, and graduate student support improved in most of the country's universities,[19] thus increasing the number of young people entering mathematics.

The Institute for Advanced Study (IAS) at Princeton is a prominent example of how the ONR benefited research. Through the support provided by the ONR, the Institute was able to expand many of its areas of research, becoming the major center for pure mathematical research that it remains today. The Institute started to receive many post-doctorates and visiting researchers during these years. These scholars then took their new expertise back to their home institutions, impacting mathematical research all over the country. Rees stated:

> Support by the Office of Naval Research for research in the more abstract fields of mathematics, the type usually represented at the Institute, had been in our planning, but authority for such support without regard to relevance to the Navy's mission had not been made explicit when the Mathematics Branch was established. Nevertheless, it seemed clear to us in the Mathematics Branch that the argument for increasing the number of well educated and experienced mathematicians was a strong one ... Moreover, there was considerable feeling among those of us responsible for the program that our concern must be with strengthening of mathematical research in the United States ... we wanted very much not to exclude any first class research.[20]

---

[19][75].
[20]*Ibid.*

Though the primary goal of the Navy through the Mathematics Branch was to increase support for students and research, and not necessarily to promote certain areas of study, there were certain fields that the ONR had great interest in promoting. Some were classified, and university researchers would be brought together for the summer to work on these projects. For example, the National Security Agency (NSA) had some problems in cryptology that they realized needed mathematical expertise. The NSA came to Rees, and she set up summer research projects to work on small parts of the problem. Rees participated in and directed several of these summer programs herself.

The non-classified areas of focus were pure mathematics, applied mathematics, statistics, logistics, analysis, and computers. Four-fifths of the budget went toward the last five categories above, and only one-fifth towards pure mathematics, however, theoretical studies made up one-third of all contracts.

The breakdown of the number of researchers in the mathematical fields with ONR contracts in 1947 was as follows:[21]

Field	Senior Researcher	Graduate Students
Applied Mathematics	36	46
Mathematical Statistics	16	26
Computing Projects	42	38
Pure Mathematics	36	10

Besides basic research, the Mathematics Branch pursued four other goals: to assist the work of research groups which were active in applied fields; to facilitate the formation of strong new research groups in these fields; to provide for the training of promising young researchers in these fields; and to make the results of this work available to the Navy and the scientific community in general.[22] A secondary hope was that by creating or enlarging research groups at universities, related courses and seminars would also be offered, increasing interaction among researchers and students. Significantly, the ONR encouraged the publication of results, publishing and even translating results when no other

---

[21][53].
[22]*Ibid.*

avenue could be found.[23] Some of the pure mathematical topics funded were the following:

- conformal mapping
- spectral theory for normal operators in function spaces
- free groups
- relationships between analytic and measure theoretical concepts of surface area with possible applications to the theory of the calculus of variations
- analytic number theory of prime number distributions and the representation of primes by polynomials
- random chemical, physical, and biological processes
- deformations in topology.

Statistical research likewise received a large boost from the ONR. Of particular importance was the work of Abraham Wald at Columbia in sequential analysis and decision theory started during the war. Work on design of experiments and data analysis was conducted by Jerzy Neyman and Samuel S. Wilks at Berkeley. Other areas promoted by the ONR were probabilistic methods, small sample theory (at Princeton), theoretical examinations of weather modification, and modeling in medicine. During the war, quality control and acceptance sampling were of high importance to the military. This focus continued after the war and in 1949 the Joint Services Program in Quality Control was established. Rees was the ONR representative to this program. ONR support of statistics continued well into the 1950s, at which point industrial funding began. Stanford University statistician Albert Bowker felt that "Mina and ONR [had] not been given enough credit for the development of mathematical statistics" in the U.S.[24] Bowker would go on to become the second Chancellor of CUNY and would work closely with Rees during her time as the GSUC.

Analysis was an area particular interest. In particular, research in analysis being conducted at New York University was supported by the ONR. This support helped NYU become prominent in the field. NYU researchers pursued questions in areas such as gas dynamics and the theory of gravity waves in

---

[23]The ONR even funded an English translation of Jakob Bernoulli's 1713 *Ars Conjectandi*. It was released as a technical report by the Harvard Department of Statistics.

[24][**30**].

water. Various other topics in mathematical physics were investigated at various institutions, such as wave propagation, two-dimensional free jets, plasticity, electrostatics, and perturbation theory of linear operators in a continuous spectrum. For example, Solomon Lefschetz set up a program in differential analysis at Princeton.

Logistics was another area of research nourished by ONR support. The work of Albert W. Tucker and his students David Gale and Harold Kuhn in linear programming and game theory had far-reaching effects on research that are still at work today. Eventually a separate Logistics Branch would be organized under Rees's Mathematical Sciences Division.

The ONR was instrumental in supporting the National Applied Mathematics Laboratory at the National Bureau of Standards. One of the important components of this program was to bring foreign mathematicians to the U.S. such as Olga Taussky Todd and Jack Todd.

Not commonly known is the fact that many engineering schools and engineering and mathematical sciences departments were expanded with ONR support. For example, Stanford University increased not only its engineering program, but also its programs in pure, applied, and statistical mathematics.

5.2. Early Computers

However important her work at the ONR was to the growth of university research in the mathematical sciences, Rees is probably most widely known for her work and influence on the early development and uses of computers.

Early work in computing was conducted as part of the war effort. By the end of the war, the speed, capacity, and applications of the new computers had grown to the point of drawing the attention of many scientists, mathematicians, and the government. Though there was controversy over whether there would ever be enough need for large computers, the Mathematics Branch of the Office of Naval Research, under Rees's direction, took the lead in much of the research into the new machines. Most of this research was conducted at universities through ONR contracts. Along with research into computing, Rees and her colleagues realized that numerical analysis would become a vital component as the machines started to be used more widely in scientific research.

Pressure from the National Bureau of Standards, the Census Bureau, and the military, pushed the U.S. government, through its various agencies, to make the United States the leader in computer research from 1946 to 1953. By 1953, industry would take the lead in research, but surprisingly not without substantial encouragement from Rees.

To facilitate the development of computers, several special-interest research groups were organized in cooperation with agencies such as the National Bureau of Standards. The National Applied Mathematics Laboratories, which contained the Institute for Numerical Analysis, was established in Washington D.C to conduct research in mathematical topics related to computers. The Computing Laboratory, the Machine Development Laboratory, and the Statistical Analysis Laboratory were also created to support those respective areas of computer research.

The Applied Mathematics Executive (later Advisory) Council was established to help organize and guide this new field of computing. It consisted of representatives from the various agencies and companies with machines on order from the National Bureau of Standards. Rees was the Navy representative. In 1947, the ONR was involved in the establishment of the Association for Computing Machinery.[25] Rees was named to the initial Executive Committee of the Association of Computing Machinery in 1947, and to the nominating committee the following year.

At first the ONR was primarily supporting the development of analog machines. These would provide two-dimensional Fourier Analysis for use in crystal structure research. Rees stated that at this time these were only projects of the ONR, and were not yet programs in their own right.[26] But very quickly, the focus turned to digital machines.

The ONR was involved with and supported, either wholly or in part, the development of several important early digital computers during Rees's eight

---

[25]The ACM grew out of the former Eastern Association for Computing [17].
[26][39].

years with the ONR. The UNIVAC of the Census Bureau and the SEAC of the National Bureau of Standards were the first large computers, both becoming operational in 1951. The SEAC was developed in part to test out the new simplex method of solving systems of linear problems that was of central importance to the Air Force. The Whirlwind computer at MIT was developed out of war work on flight simulators and focused on real-time operation. The von Neumann-Bigelow-Goldstine machine was developed at the Institute for Advanced Study.[27] The Institute for Numerical Analysis constructed the SWAC computer. ONR also supported a machine at Pennsylvania State College to analyze crystallographic structures. While the Logistics Research Project built the ONR Logistics Computer at George Washington University. Rees also supported the development of simple and cheap computers[28] to be used widely at universities, again against stiff resistance from industry, which maintained that only a handful of large machines were needed.

The ONR supported research in computing in order to facilitate advances in other fields, such as weather and flood forecasting, hydrodynamics, wave theory, explosives, optics, radiation, and atomic power. Interestingly, the SEAC computer at the Bureau of Standards was used to compute the lowest bidder on government contracts. This problem used complex systems of equations requiring linear programming. This promoted new applications of mathematics to government and business situations, and the development of new mathematical fields. Similarly, the ONR and the Bureau of Standards created a patent-searching machine for the U.S. Patent Office.

Possibly Rees's greatest contribution to computing and to the support of mathematics was her realization that computing could not move forward without solid and comprehensive mathematical backing. She believed that, "It is necessary, not only to design machines for the mathematics, but also to develop

---

[27][3]. By 1954, the methods developed by John von Neumann at Institute for Advanced Study for weather forecasting were being used by all three of the U.S. military branches and the U.S. Weather Bureau.

[28]This is a relative statement. The 'simple and cheap' machines Rees foresaw were the mainframe machines common in universities and school districts across the country in the 1980s. With the advent of PCs, these mainframes, especially in public schools, virtually disappeared by the turn of the century.

new mathematics for the machines."[29] She gave support to many mathematical projects, either directly or indirectly related to the work on computers. In
particular, she supported a revival in numerical analysis, needed in the development and use of the new digital computers. This was done by supporting a
program at the Institute for Advanced Study, and by establishing the Institute
for Numerical Analysis at UCLA under the direction of John von Neumann.
The Institute for Numerical Analysis investigated how numerical analysis could
be used to enhance the design of the new computers, and examined mathematical problems that arose during the construction and use of the computers.

Through reading the various articles Rees wrote describing the development
of computers, it becomes clear that not only did Rees have a wide knowledge
of computing, but a keen insight into the future innovations in computing.
Through her travels and interactions with researchers across the country, Rees
gained a unique perspective that few if any others had that allowed her to see
various pieces of the puzzle, and perhaps see the future of computing more
clearly. For example, she was an avid proponent of magnetic core memory and
later electrostatic memory. She foresaw the switch from vacuum tubes to tran-
sistors, when many people in the field did not. She pushed for faster and faster
machines and with larger memories, foreseeing the commercial and academic
uses of computers to come, and strove to involve industry in the development
of the new machines, again surprisingly against strong resistance. "As I left
Washington [in 1953] I was convinced that computers would be exploited in
the future by the large commercial companies and the role of the government
vis-a-vis the computer would be quite different from what it had been during
the zesty days of my service in the Office of Naval Research."[30] She promoted
multiple inputs and visual displays for output (today's monitors). (Up to this
point, all output was printed on paper similar to a ticker tape.) As an interesting point, Rees was instrumental in the Atomic Energy Commission's decision
to place the UNIVAC 4 computer at NYU; this would be vital to the development of the Courant Institute at NYU, which was founded in 1953.[31] Henry
Tropp commented that, "[i]n her job at the Office of Naval Research, [Rees]
had up-to-date information on what was going on globally. By virtue of the
demands on ONR and the role of the federal government in both funding and

---

[29][**55**, p. 335].
[30][**79**, p. 119].
[31][**26**].

anticipating future computational needs, she was in a position to know what was being done, what was working, and what 'future' technology anticipated."[32]

One example of the breadth of support provided by the ONR to assure the free flow of information and ideas, as well as Rees's pervasive presence in all aspects of the research program, is the January 1947 "Symposium of Large Scale Digital Calculating Machinery." This four-day conference held at Harvard and the Massachusetts Institute of Technology was sponsored jointly by Harvard, the U.S. Bureau of Ordnance and the Department of the Navy (in coordination with the ONR). Rees chaired a session on "Numerical Methods and Suggested Problems for Solution" (in which Richard Courant spoke). Rees also gave the Friday evening banquet address, speaking on "Future Field of Application of High-speed Computers."[33]

Rees took a short leave from the ONR for a sabbatical at NYU in 1952. The result of this sabbatical was the publication of "On the Solution of Nonlinear Hyperbolic Differential Equations by Finite Differences," with Richard Courant and Eugene Isaacson.[34] Round-off and truncation errors along with enhancing the efficiency of numerical calculations being used in the new computers was a prime concern. In particular they wanted to reduce errors that occurred when continuous problems were solved using discrete methods. Their paper focused on quasi-linear hyperbolic equations. To do this they compared several methods for calculating differences. They found that using a rectangular net with specific conditions on mesh width was best adapted to computer usage. This sabbatical and the publication of a joint paper show that Rees had not completely given up her interest in pursuing pure research.

This was the only return to pure research Rees would make. She found that once out of mathematics it was hard to get back into it. She also felt that her talents lay much more in the world of administration, and that policy work took up all of her time. "I think that [not doing mathematics] has been a deprivation for me; I have lost touch with things I cared about in mathematics

---

[32][**80**, p. 156].
[33][**8**].
[34][**7**].

...I don't think it's possible once you have left mathematics to get back into it."[35]

Rees stated, "In the early days, we recognized that, until a National Science Foundation (NSF) was established, ONR had a special obligation to provide for the balanced support and growth of mathematical research in the United States, always, of course, within the framework of the Navy's established policy."[36] To date, the ONR has funded fifty-eight Nobel laureates.[37] Rees and her staff at the ONR lent much support and advice during the establishment of the NSF in 1950.

≪   ≫

Overall, Rees felt that her biggest contribution while in a policy position in Washington was to make sure that mathematics got a fair share of support. She did this by continuously demonstrating the achievements of mathematics, and by being continually aware of opportunities to enhance mathematical research.

Rees received many awards and citations for her work in the Office of Naval Research and her impact on mathematical research. She received the Public Welfare Medal of the National Academy of Sciences in 1983 for 'eminence in the application of science to the public welfare'. However, the honors she received from mathematical groups were the most meaningful to her. She received the Mathematical Association of America's first Award for Distinguished Service to Mathematics in 1962. This award is given for outstanding service to mathematics, other than mathematical research. The American Mathematical Society and the Institute of Mathematical Statisticians both adopted resolutions in honor of her. An excerpt from the latter reads as follows, "Under Dr. Rees' leadership the Division of Mathematical Sciences of the Office of Naval Research gave whole-hearted support to basic research ... The whole action was conducted with remarkable foresight and wisdom ... Mathematical Statistics owes Mina Rees a public 'well done'."[38]

---

[35][9].
[36][75].
[37][43].
[38][77, p. 17].

As this citation shows, Rees's value lay in her leadership abilities. "[T]he importance of the ONR experience was in the rather intimate knowledge it gave me of the *modus operandi* and the ambiance of virtually all of the country's leading research universities and of many of the liberal arts colleges. It also resulted in the establishment of warm friendships with many mathematicians and with many university administrators."[39] From her years on the Applied Mathematics Panel and her travels before joining ONR, she developed a broad sense of what mathematical research was and how best to facilitate it. She also developed a keen sense of the future of research. Through her insight and care, Rees shaped the post-war scientific research community in the United States, and gave us the form of research support that we take for granted today.

These experiences became invaluable to Rees when she returned to Hunter College to begin yet another phase in her career, that of academic administration. After leaving Washington, Rees still remained involved with the work promoted by the ONR. For example, in 1953 she became a member of the Advisory Group established by Richard Courant at the time the Institute for Mathematics and Mechanics at NYU became the Courant Mathematics and Computing Laboratory (usually referred to as the Courant Institute). She was also a liaison for the Navy. While she was Dean of Graduate Studies at CUNY, the Navy sent officers pursuing Ph.D.'s to Rees to aid them in their transition from the military into graduate school, both at CUNY and at other institutions.

---

[39] *Ibid.*

CHAPTER 6

Rees and Graduate Education: 1953–1972

Can we have excellence and equality [in education] or must we choose between them?
– Mina Rees
The Ivory Tower and the Marketplace, 1976

Rees spent the final phase of her career back in New York City as a professor of mathematics and an administrator. Her greatest impact during this time was upon graduate education – specifically, at the City University of New York, but more generally, throughout the United States.

6.1. 1953–1972

The Council of the American Mathematical Society here takes cognizance of the resignation this past September of Dr. Mina Rees as Head of the Mathematical Section of the Office of Naval Research. She has accepted a position as Dean of the Faculty at Hunter College. We congratulate Hunter College on this wise selection and can only say that our heavy loss as mathematicians is the gain of Hunter College.
– Executive Council of the AMS, December 29, 1953, Baltimore, Maryland[1]

Rees returned to New York in 1953 after leaving Washington D.C. and the Office of Naval Research to rejoin the faculty of Hunter College. She served as Professor of Mathematics and Dean of Faculty at Hunter for eight years.

---

[1][36].

In 1954 she was appointed to launch and head Hunter's Office of Institutional Research.

While Dean of Faculty at Hunter, Rees strove to make her alma mater a center of excellence in undergraduate education. F. Joachim Weyl describes her presence at Hunter during this time:

> When, for the third time, she settled at Hunter College, as Dean of Faculty, it was not only as a talented but as a seasoned administrator whose decisions were clear, timely, and – if not always popular – certainly respected. The College was evidently both pleased and surprised to see what had become of their Mina Rees, but at times – if some of the long-timers have been rightly interpreted – it appeared to be a bit more surprised than pleased. Stimulated by the resources on hand and the opportunities available, she set Hunter to moving faster and more purposefully toward programs that reached higher up the ladder of academic education and responded to a wider range of the city's needs.[2]

When the City University of New York was established in 1961, she became the first Dean of Graduate Studies. In this role, and throughout the remainder of her career at CUNY, she was responsible for building the graduate program of CUNY. In 1968, she was appointed Provost of the Graduate Division and in 1969 became the first President of the Graduate Division. She became President of the newly established Graduate School and University Center of CUNY that same year, and retired in 1972 as President Emeritus.

When Rees moved to oversee graduate studies at CUNY, she continued to strive for excellence and to promote diversity in the graduate programs of the University. During her years at CUNY, Rees impacted not only graduate education at CUNY, but graduate education in general, and the graduate education of women in particular. Beyond CUNY, during this same time, Rees was an active member of the Council of Graduate Schools in the United State. So Rees was able to affect profound change not only at the local, but at the national level as well.

---

[2][91, p. 1150].

She became involved in education and educational reform at all levels, including secondary education. For example, in 1958 she was asked by the NSF to organize a meeting of educators to address concerns over mathematics education at the elementary and secondary level. The meeting was held at MIT, and the result was the influential School Mathematics Study Group. The Group published several texts and pamphlets that outlined broad as well as specific reforms in school mathematics. The most notable of the reforms was the "new math" of the 1960s that energetically promoted the use of set theory in the teaching of mathematics at the elementary school level. In the early 1960s Rees was a member of the advisory panel of the School Mathematics Study Group.

Rees was also a member of several committees and panels focusing on curriculum guidelines in the sciences and mathematics, such as the Conference Board of the Mathematical Sciences and the Executive Committee of the Mathematical Division of the National Research Council. She served on many committees and panels that addressed the issues and educational needs of young people entering mathematics at the high school, college, and graduate levels, as well as those entering the job market in mathematics. She published several articles addressing these issues.[3] These articles described the type of mathematical background a student would need in the future, as our society became more technical. Rees also served on several committees that dealt with undergraduate and graduate education in general, such as the Committee on the Undergraduate Program of the Mathematical Association of America. Rees chaired the Non-teaching Opportunities Committee for the Survey of Research Potential and Training in the Mathematical Sciences, funded by the National Science Foundation in 1954.[4]

Besides publishing articles and reports in education, Rees published several articles of a mathematical nature dealing with such topics as the history of mathematics, philosophy, population control, environmental issues, the development of computers, and the role of science and scientists in society. Rees was also a sought after speaker.

Rees maintained her involvement in government policy issues by being a member of several committees including the Advisory Panel for Mathematics

---

[3]See entries in the bibliography for Rees from 1951 to 1965.
[4]See Appendix J for a list of her contributions.

of the National Science Foundation, and the General Science Advisory Panel of the Department of Defense, to name just two.[5] President Lyndon Johnson appointed her to the National Science Board in 1964. In 1971 Rees became the first female President of the American Association for the Advancement of Science. Lynn Osen states in *Women in Mathematics* that " ... future historians will surely record the illustrious career of such women as Mina Rees who has the honor of being the first woman president of the nation's largest scientific society, the Association for the Advancement of Science."[6] An editorial in the New York Times at the time of her election claimed, "The examples set by Marie Curie, Lisa Meitner, Margaret Mead, Mina Rees and many others prove that scientific creativity is not a male monopoly."[7] As President, Rees's goal was to encourage the U.S. government to continue to lead the way in funding scientific and technological research. Rees had been Vice President and Chairman of Section A of the association in 1953–1954, and was later a member of the Board of Directors.

6.2. Rees's Impact on Graduate Education

> Dr. Rees, as the only woman dean of a graduate school in a co-educational institution, has developed an entirely new career for herself. And she is one of the few deans of graduate instruction who are pioneering with a new concept of how graduate instruction can be kept at a superior level under today's many and increasing pressures.
>
> – Juanita Massa, AAUW Fellowships Chairman, 1965[8]

The City University of New York was organized in 1961, consisting of the five senior colleges of New York City: City College, Hunter College, Brooklyn College, Queens College, and Lehman College, along with some of the city's junior colleges. Rees commented that CUNY "resembled a British University with its many largely autonomous colleges"[9] One of the purposes of the new City University of New York was to provide graduate education. The first Chancellor

---

[5]*Ibid.*
[6][**44**, p. 154].
[7][**25**].
[8][**66**, p. 59].
[9][**77**, p. 18].

of CUNY, John Everett, appointed Rees as the first Dean of Graduate Studies, as well as to the Board of Higher Education for the CUNY system.[10] Her job was to develop a new graduate program as fast as possible that would be competitive with the top graduate schools in the nation, and cater to the large and diverse student body of New York City. Rees stated that, "The building of a graduate school that called upon and stimulated the growth of the scholarly and physical resources of so many established liberal arts colleges and that achieved acknowledged first-class graduate work in a brief period of time required the combining of traditional elements of academic structure with often difficult innovations."[11]

Her challenge would be to bring together a high quality faculty to serve both the new graduate programs and the undergraduate colleges, design the graduate programs, and to create new facilities, all without duplication. She was involved with every aspect of this enormous task from hiring faculty to decorating the new facilities. Joan Byers, Rees's secretary from 1961 to 1966 and her Special Assistant in 1971, worked closely with Rees during the crucial years when the graduate school was taking shape. Byers recalls that Rees was a tireless worker;[12] and that Rees not only pushed herself to do the best job possible, but also encouraged and pushed those around her to excellence.[13]

Originally the graduate program had only four programs: chemistry, economics, English, and psychology. In 1969, the Graduate School and University Center (GSUC) was established in Manhattan to house the graduate programs in the social sciences, humanities, mathematics, education, and non-laboratory activities. By 1971, GSUC had twenty-six doctoral programs. During the early years of CUNY, programs were housed throughout the city. Rees was appointed President by then Chancellor, Albert Bowker. As President, she constructed GSUC on the consortium model. She set the standards for the institution,

---

[10][5].

[11]*Ibid.*, p. 18.

[12]*Ibid.*

[13]Each year Rees issued an Annual Report of the Dean of Graduate Studies, dealing with her concerns, goals, and projects at GSUC. In 1982 Rees published *The First Ten Years of the Graduate School: The City University of New York.* This work described the progress of the Graduate School in its early years.

which encompassed all facets of the graduate program.[14] Frances D. Horowitz, President of GSUC at the time of Rees's death stated in a memorial that, "Mina Rees's understanding of what it means to do quality graduate education set the founding base for the City University Graduate School – a base upon which it was possible to build to the eminence that the GSUC now enjoys."[15]

Overall, Rees strove to create a graduate program at CUNY that kept pace with an increasingly changing society. Rees advocated graduate programs that were flexible for the student. In particular she encouraged internships, possibly several, in different areas of one's field. She felt that in certain cases, time spent working in the field for an extended period before completion of a degree was not only beneficial, but necessary. She also felt it was important for faculty members to return to industry occasionally, to keep up to date and to encourage interaction between academia and industry. She was particularly interested in problem-oriented courses of study.[16] Rees hoped to develop this kind of interaction between institutions also. She organized cooperative programs with institutions such as the American Museum of Natural History and New York University. She believed that, "[t]he attempt to achieve distinction rather than minimum standards must rely, I believe, on the professional identification of faculty members and on national cooperation within a discipline."[17] She also felt that there was ample need for a degree at the Ph.D. level but without the need for the dissertation (what we would now call ABD) for persons desiring to work in industry.[18]

Rees designed the graduate programs at CUNY with the non-traditional student body of New York in mind. Consequently she developed a large program of remedial study at CUNY. She knew that students entering graduate school came from vastly different social and educational backgrounds, and may be returning to academia after a prolonged absence. She was determined to

---

[14]Note that as President, Rees oversaw all graduate programs, not just those in mathematics or the sciences.

[15][25].

[16]Rees advocated these ideas over a generation ago. Yet programs with these types of options are toted as being cutting edge today.

[17][67, p. 228]. And again, we hear arguments today deploring the results of focusing on minimum standards to the detriment of high quality programs.

[18]The author obtained a D.A. (Doctor of Arts) degree, as opposed to a Ph.D., in one of the few remaining such programs offered in the U.S.

give every student the chance to succeed. "It is important, if our total system of higher education is to meet its obligations, that each campus assess
where its best contributions will lie in the light of its history, its resources, its
commitments, and the community it serves."[19]

Rees was also a strong proponent of the idea that good teachers must be
good researchers, and that students should be exposed to research in their field
as early as possible. Thus faculty taught both graduate and undergraduate
courses. Rees stated,

> [T]o create a university without securing the strength of the
> undergraduate education on which it rests is to ignore one of
> the country's most pressing educational problems. To try to
> strengthen the colleges in the United States without some at-
> tention to the lure of research for able scholar-teachers is to
> ignore a central reality in this area.[20]

With acute insight, Rees saw that as women's and minorities' roles changed
in this country, the family would change. With the change in traditional roles,
graduate eduction would have to adapt in order to serve as many students as
possible.[21] Rees was particularly conscious of women graduate students and
their needs. She devoted much time and energy to bettering graduate education for women. She promoted part time employment and study for female
students with children, along with day-care centers. She considered the need for
a babysitter an appropriate reason for financial support. She encouraged graduate schools to consider programs that would help women to update their knowledge upon returning to graduate work after raising a family. Possibly above all,
Rees believed in providing strong role models for women, knocking tradition
when necessary, " ... we may see here a self-fulfilling prophecy if women are
denied the opportunity to qualify for leadership positions in academia because,

---

[19][**74**, p. 99].

[20][**66**, p. 34].

[21]Even today, most graduate schools are organized on a traditional model and that
limits the ability of non-traditional students to attend. Though there are many more evening
programs than previously, most of these are not in the hard sciences. I think the recent advent
of on-line degree programs in many fields would fit in nicely with Rees's vision, as well as her
early views on the future use of computers.

in the past, they have not been appointed to them."[22] She wrote several papers on the graduate education of women[23], directed both at female students and at administrators. She was a member of several committees that considered women's issues related to graduate education.

In a 1972 paper, Rees stated, "[A]t the City University, in contrast to the situation reported somewhat earlier at several other universities, there was strong evidence that women's performance as graduate students was about the same as that of men with respect to all three parameters that had been used to measure this performance in other studies reported in the literature: completion of the first or qualifying examinations, completion of all requirements for the degree except the dissertation, and completion of the degree."[24]

While reading the various articles concerning graduate education in the United States that Rees wrote, it becomes clear that Rees had very high and noble expectations for the future of graduate education. She had a view of graduate programs that would foster independent thinking on the part of the student, provide a healthy and supportive research environment for the faculty, and that were flexible and adjustable to the needs of the student. She believed in an atmosphere of open cooperation between industry and academia. Rees challenged the graduate community to open discussions on such controversial issues as open admissions and alternative programs of study for non-traditional students. Above all, Rees envisioned a future for academia in which the solving of social issues was of prime importance, to every student, faculty member, and department. She hoped to bring academics out of the 'ivory tower', and into valuable contact with the issues of the day.

To acquire a better understanding of her views on these matters, let us consider her own words. From *The Ivory Tower and the Market Place*:[25]

---

[22][**72**, p. 185].

[23][**65**] and [**72**].

[24][**82**, p. 18]. The first two recipients of the Ph.D. at the GSUC were women, one in English and one in psychology.

[25][**74**].

The demands of society challenge us to lend the vitality of the
ivory tower to the solution of the much more difficult problems of the marketplace, where a jump must be made from the
security of knowledge to the insecurity of decision.
[...]
I would hold that a continuing major function of the university
in the United States is to support and strengthen the kind
of inquiry into nature which can be pursued chiefly from a
motivation to understand the world and the people in it.
[...]
[I]t is important to recognize that in this vast and varied country of ours there are many functions to be served and diverse
clienteles to be educated, and it would serve us ill if all our
universities set for themselves the same purposes, the same
organization patterns, and the same campus ambience.
[...]
In the commitment to equal access [to education] lies the decisive component to the [current] transition ... in the history of
America's universities.
[...]
Can we have excellence and equality or must we choose between
them?

From *Graduate Education – A Long Look*:[26]

What, indeed, is graduate education for? Is it for getting a job?
For providing the understanding out of which criticism of our
social goals and national actions can be built? Is it for pushing
back the boundaries of knowledge and decreasing the areas of
ignorance? Or is it, perhaps, for the glory of the human spirit?
Is it for the student? For the Faculty? Or for society?
[...]
There should, I think, be a number of great population and
cultural centers in the country in which pure scholarship, in all
fields, is pursued, although not necessarily at a single university
nor even exclusively at universities.

---

[26][**70**].

[...]
Another of the triumphs of American Ph.D. education has been
the success of the graduate schools in putting out many workers
in the vineyard who have built the base on which the work of
the great researchers has stood.

The middle part of the quote above clearly is a result of Rees's experiences
on the Applied Mathematics Panel and at the ONR.

≪ ≫

Rees worked tirelessly to increase the effectiveness and scope of graduate
education in the United States. She added to the educational opportunities
of women and minorities. She pushed herself and those around her towards
excellence. She always looked for new perspectives. Lynn Osen felt that, "[Rees]
succeeded because she expected to succeed and demanded no less from herself
and those around her."[27]

Rees's work at CUNY led to her receiving New York City Mayors award in
science and technology twice, in 1964 from Mayor Robert Wagner and in 1984
from Mayor Ed Koch.

The graduate program at the City University of New York was her testing
ground. Thus her efforts on the national level were practical, timely, and informed. Through her words and presence, she forced her colleagues to look to
the future of education, and not to stick with tradition. She was not hindered
by criticism or norms.

After her retirement, the Graduate School and City Center inaugurated the
Mina Rees Dissertation Fellowship for students finishing their degree. Hunter
College also established a fellowship in her honor. In her will, Rees provided $1.7
million for GSUC to establish a chair in mathematics in her name.[28] In 1985,

---

[27][**45**, p. 179].
[28]There was some local uproar in New York in 2002 when the chair was awarded to
Russian (as opposed to American) mathematician Victor A. Kolyvagin, who still holds the
chair.

GSUC further honored Rees by renaming its library the Mina Rees Library "in tribute to her remarkable qualities as administrator, teacher, and colleague, and in recognition of her critical role in establishing doctoral work at The City University of New York."

For many years after her official retirement, Rees continued to be involved with education and research by continuing to serve on many of the committees and boards that she had contributed to over the years. She received nineteen honorary degrees, and was asked to give commencement addresses at many institutions. As with her earlier work during and following the war, Rees's presence in graduate education will be felt for years, and even generations to come.

FIGURE 6.1. National Science Board, February 13, 1969. Rees second on right. Courtesy of the Archives of the Graduate Center, CUNY.

FIGURE 6.2. Mina Rees with President Lyndon B. Johnson and his wife Claudia "Lady Bird" Johnson at the White House in 1965 to accept her appointment to the National Science Board of the National Science Foundation. The portrait in the background is of President James Monroe. Courtesy of the Archives of the Graduate Center, CUNY.

FIGURE 6.3. Signed photo of Mayor Koch of New York City awarding Rees the Mayor's Award for Science and Technology, 1986. Courtsey of the Archives of the Graduate Center, CUNY.

FIGURE 6.4. Mina Rees and Leoplod "Lee" Brahdy in their apartment in Manhattan, c. 1970. Courtesy of the Archives of the Graduate Center, CUNY.

CHAPTER 7

Conclusion

The National Academy of Sciences' Public Welfare Medal is one of the most prestigious honors of the Academy. It is awarded "for distinguished contributions to the application of science for the public welfare. Rees received this distinguished award in 1983 for her work in furthering mathematics and computer science since World War II.

The Public Welfare Medal was first awarded in 1913 jointly to George Goethals for the construction of the Panama Canal, and to William Gorgas for eradicating yellow fever in the canal region. Other recipients of the award included J. Edgar Hoover and John D. Rockefeller. To place Rees in the ranks of these men, who vastly changed the American landscape, is nothing less than impressive.[1] But the first female President of the nation's most prestigious scholarly assembly must, by definition, be of outstanding character. In the 1971 article in *Science* announcing Rees's presidency of the AAAS, Weyl stated,

> Mina Rees' characteristically spontaneous way of lighting up in lucid understanding and her clarity in communicating earned her in time the warm friendship of many, and the acquaintance, vividly remembered, of most of her contemporaries in the field ... her impact on the history of mathematics does not lie in what she has added to mathematics insight but lies in the changes she has wrought in the professional lives of many significant mathematicians of her time.[2]

---

[1]See appendix G for the photograph and accompanying letter regarding the Public Welfare Medal ceremony.

[2][**91**, p. 1149].

Rees is most widely known for her policy work at the Office of Naval Research and her contributions to the development of computers during the 1950s. However, her influence, both official and personal, affected almost every corner of education as well as government policies on the support of scientific research. Lynn Olsen declared that, "Support of research enterprises in the midst of the hot and cold wars of the mid-twentieth century represented a major and meaningful demonstration of Rees' ability to balance the theoretical and practical aspects of science."[3]

One of Rees's lesser known achievements, but equally important, was in bringing mathematics to the forefront of the sciences. As Rees stated in 1985: "When I was in Washington, I felt that the most important thing I did was to see to it that mathematics got its share of support by repeatedly demonstrating its achievements. I got the support by being on hand all the time and being watchful."[4].

Rees accomplished this not only through her knowledge of the mathematical community and her administrative expertise, but through her force of will, determination, and character. However, she gives some of the credit to the mystique of mathematics. When asked in an interview if being a mathematician was useful to a person in public affairs, she had this to say. "In dealing with academics, it is absolutely superb to be able to say you're a mathematician! Nobody dares to say mathematics is not important or not significant. I have always found it was an advantage when I was dean or president of a college."[5]

Few if any mathematicians are household names. But an indication of Rees's influence, not just in the areas of mathematics, research and education addressed here, but to the wider public is that she was well enough known to be included in an expose in the June 1970 issue of Vogue magazine entitled "Liberated, All Liberated" that showcased such women as Liza Minnelli and Patrician Nixon.[6]

---

[3][**45**, p. 177].

[4][**9**, p. 260].

[5]*Ibid.*, p. 260.

[6]Another indicator of Rees's impact: a colleague of the author sent a crossword puzzle from a crossword puzzle book that had the clue: Dr Mina _____, US mathematician.

Arthur Schlesinger, Jr. remembered Rees in the GSUC memorial for her. "I well remember my first meeting with Mina – a tall, elegant lady, urbane and charming in manner but clear and purposeful in mind, *suaviter in modo, fortiter in re*.[7] As a humanist, I naturally looked on a distinguished mathematician with suspicion. But, as I quickly discovered, if there was anyone who transcended the gap between what C. P. Snow had famously called the two cultures, it was Mina Rees."[8]

Mina Rees was a truly inspiring woman. She was a benefit to mathematics and education, whose scope of influence is hard to grasp.

> If I have given the impression that all these activities are carried on by a person of engaging warmth and liveliness, of boundless energy, and – foremost of an extraordinary directness and clarity in interpretation, giving sight to imagination, then I have been understood correctly.
> -Weyl, *Science,* 1970.[9]

Suaviter in modo, Fortiter in re

---

[7]Gentle in manner, Strong in reason.
[8][**25**].
[9][**91**, p. 1151].

Course Work at Columbia University

Summer, 1923

> Advanced Course in the Teaching and Supervising of Mathematics
> Differential Geometry

Winter, 1923–1924

> History of Mathematics
> Introduction to Mathematical Philosophy

Spring, 1924

> History of Mathematics
> Theory of Infinite Series

Summer, 1924

> Philosophy of Education
> Organization and Administration of Extra Curricular Activities in
> Secondary Schools
> The Great Systems of Philosophy

Winter, 1924–1925

> Philosophy of Education
> Philosophic Aspects – Education and Nationalism

A. COURSE WORK AT COLUMBIA UNIVERSITY

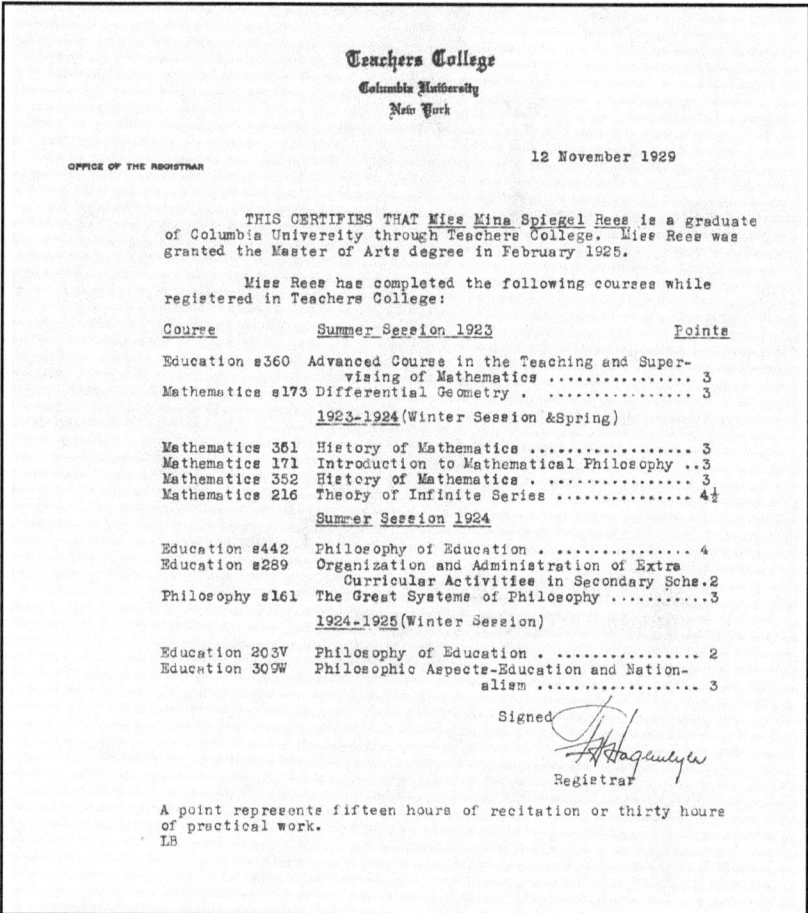

FIGURE A.1. Original transcript from Columbia University. Courtesy of the Columbia University Archives.

Course Work at the University Of Chicago

Summer 1929[1]

Elementary Theory of Equations (Raymond Barnard)
Evaluation of Real Variables
Structures of Complex Variables I

Fall 1929

Theory of Numbers (Leonard Dickson)
Structure of Functions of a Complex Variables (Mayme Logsdon)
Vector Analysis (Arthur Lunn)
Theoretical Mechanics I (Wm. Duncan MacMillian, Astronomy Department)

Winter 1930

Metric Differential Geometry (Ernest Lane)
Hermetian Matrices of Positive Type in General Analysis I (Raymond Barnard)
Theory of Functions of a Real Variable (Lawrence Graves)
Theoretical Mechanics II (Wm. Duncan MacMillian MacMillian)

Spring 1930

---

[1]Material provided to the author by the University of Chicago Office of the Registrar.

Advanced Topics in Algebra and Theory of numbers I (Leonard Dickson)

Hermetian Matrices of Positive Type in General Analysis II (Raymond Barnard)

Theoretical Mechanics III (Wm. Duncan MacMillian MacMillian)

Summer 1930

Differential Equations (Herbert Ellsworth Slaught)

Higher Plane Curves (Ernest Lane)

Theory of Numbers and Linear Algebra (Leonard Dickson)

Fall 1930

Lattices and Crystal Groups (Arthur Lunn)

Calculus of Variations I (Gilbert Bliss)

Projective Differential Geometry (Ernest Lane)

Thesis work in Theory of Numbers and Linear Algebra (Leonard Dickson)

Winter 1931

Calculus of Variations II

APPENDIX C

King's Medal of Great Britain

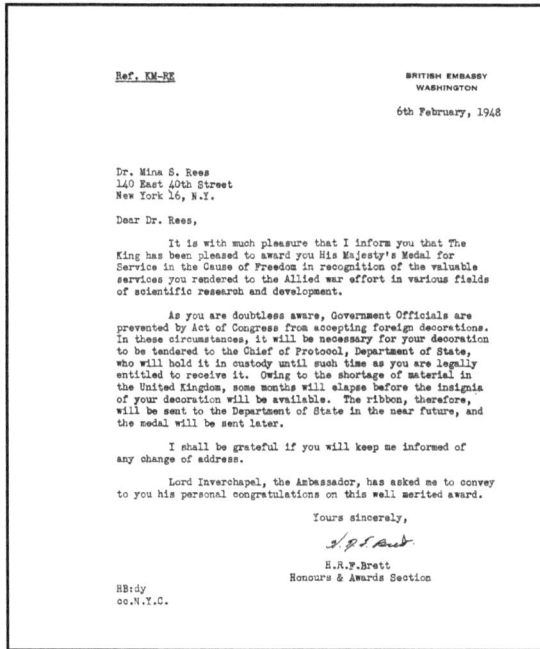

Ref. KM-RE

BRITISH EMBASSY
WASHINGTON

6th February, 1948

Dr. Mina S. Rees
140 East 40th Street
New York 16, N.Y.

Dear Dr. Rees,

It is with much pleasure that I inform you that The King has been pleased to award you His Majesty's Medal for Service in the Cause of Freedom in recognition of the valuable services you rendered to the Allied war effort in various fields of scientific research and development.

As you are doubtless aware, Government Officials are prevented by Act of Congress from accepting foreign decorations. In these circumstances, it will be necessary for your decoration to be tendered to the Chief of Protocol, Department of State, who will hold it in custody until such time as you are legally entitled to receive it. Owing to the shortage of material in the United Kingdom, some months will elapse before the insignia of your decoration will be available. The ribbon, therefore, will be sent to the Department of State in the near future, and the medal will be sent later.

I shall be grateful if you will keep me informed of any change of address.

Lord Inverchapel, the Ambassador, has asked me to convey to you his personal congratulations on this well merited award.

Yours sincerely,

H.R.F.Brett
Honours & Awards Section

HB:dy
cc.N.Y.C.

FIGURE C.1. King's Medal of Great Britain, awarded to Rees by King George VI , 1948. Courtesy of the Archives of the Graduate Center, CUNY.

APPENDIX D

Presidential Certificate of Merit

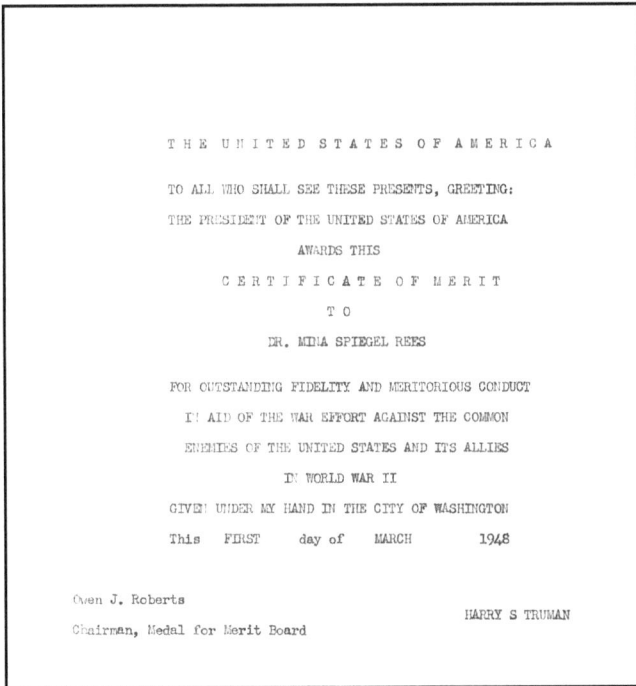

THE UNITED STATES OF AMERICA

TO ALL WHO SHALL SEE THESE PRESENTS, GREETING:

THE PRESIDENT OF THE UNITED STATES OF AMERICA

AWARDS THIS

CERTIFICATE OF MERIT

TO

DR. MINA SPIEGEL REES

FOR OUTSTANDING FIDELITY AND MERITORIOUS CONDUCT

IN AID OF THE WAR EFFORT AGAINST THE COMMON

ENEMIES OF THE UNITED STATES AND ITS ALLIES

IN WORLD WAR II

GIVEN UNDER MY HAND IN THE CITY OF WASHINGTON

This   FIRST   day of   MARCH   1948

Owen J. Roberts

Chairman, Medal for Merit Board

HARRY S TRUMAN

FIGURE D.1. Presidential Certificate of Merit awarded to Rees by President Truman in 1948. Courtesy of the Archives of the Graduate Center, CUNY.

MAA Award for Distinguished Service to Mathematics

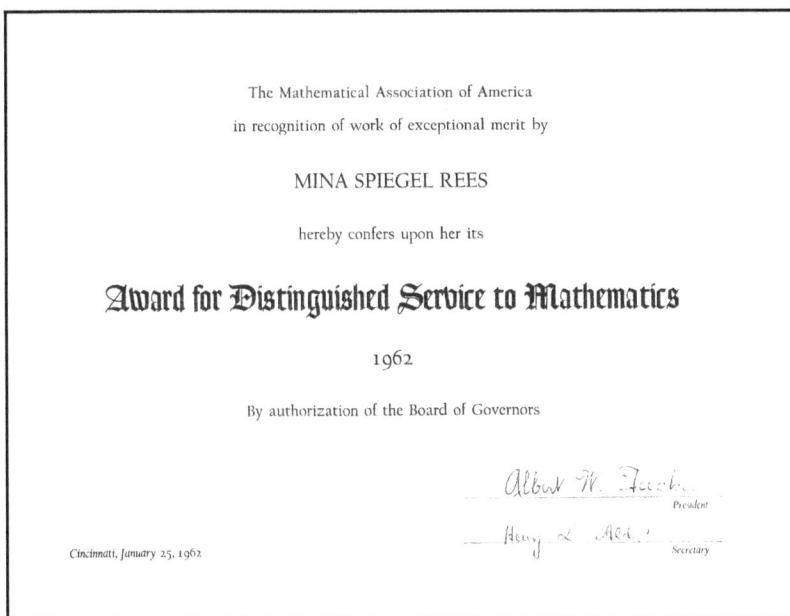

The Mathematical Association of America

in recognition of work of exceptional merit by

MINA SPIEGEL REES

hereby confers upon her its

Award for Distinguished Service to Mathematics

1962

By authorization of the Board of Governors

Albert W. Tucker
President

Harry ... ...
Secretary

Cincinnati, January 25, 1962

FIGURE E.1. MAA Award Certificate. Courtesy of the Archives of the Graduate Center of CUNY.

THE MATHEMATICAL ASSOCIATION OF AMERICA

Official Reports and Communications

AWARD FOR DISTINGUISHED SERVICE TO MATHEMATICS

The Board of Governors of the Mathematical Association of America at its meeting on August 28, 1961, in Stillwater, Oklahoma, has voted to name Dr. Mina S. Rees (Mrs. L. Brahdy), Dean of Graduate Studies of the City University of New York, as the first recipient of the Award for Distinguished Service to Mathematics.

A certificate and monetary award in the amount of five hundred dollars was presented to Dean Rees at the time of the January 1962 meeting of the Association in Cincinnati, Ohio.

In establishing the conditions of this recognition it was specified that the award was to be made "for outstanding service to mathematics, other than mathematical research" and the "the contribution should be such as to influence significantly the field of mathematics or mathematical education on a national scale" (not necessarily in the United States).

The following citation was prepared by Professor Wallace Givens, Chairman of the Committee on the Award for Distinguished Service to Mathematics, with the assistance of information contributed by Dr. F. J. Weyl and President John J. Meng of Hunter College.

Citation of Mina S. Rees (Mrs. L. Brahdy).

The distinguished career of Dean Mina S. Rees in administration of government and academic enterprises which is here recognized was based on a sound mathematical education at Hunter College, where she graduated summa cum laude, at Columbia University and at the University of Chicago which awarded her the degree of Doctor of Philosophy in 1931. After teaching mathematics at Hunter College as Instructor, Assistant Professor and Associate Professor during the years 1926-43, she entered government service in 1943 as Technical Aide and Executive Assistant to the Chief of the Applied Mathematics Panel of the

Office of Scientific Research and Development. In 1946 she was named Head of the Mathematics Branch of the Office of Navel Research and later (1949-53) as Deputy Science Director. During these years in Washington, Mina Rees firmly built into the permanent structure and policies of the ONR the principle that the full scope of mathematics should form part of the total scientific effort properly supported by government sponsored research programs.

In recognition of her service during the war years, Dean Rees was awarded the President's Certificate of Merit as well as the King's Medal for Service in the Cause of Freedom from Britain.

Of her work in establishing the current pattern of research in mathematics, a Washington colleague has expressed his appreciation in the following terms:

"Through her personal imagination, initiative, and judgement, as well as her indefatigable leadership in rousing this county's mathematical community to state its case and turn its effort to the development of ways for the effective use of the research support funds which thus became available to it in increasing amounts, she has contributed perhaps more than any other single person to the scope and wealth of present day mathematical research activity in the United States which has been made possible by the judiciously channeled massive injection of Federal funds."

Resolution of the Council of the AMS

The Council of the American Mathematical Society here takes cognizance of the resignation this past September of Dr. Mina Rees as Head of the Mathematics Section of the Office of Naval Research. She has accepted a position as Dean of the Faculty at Hunter College. We congratulate Hunter College on this wise selection and can only say that our heavy loss as mathematicians is the gain of Hunter College.

The very striking and brilliant contributions made by pure (non-military, non-applied) science, not least of these by mathematicians, to the winning of World War II is well known. It was clearly seen by the government and those responsible for the armed services that a large scale fostering by the U.S. government of fundamental research, the basis of all research, was unavoidable. Only thus could we hope to hold our own in years to come, and incidentally build up a suitable reserve of talented men for emergencies. This was actually acted upon by the Navy who thus took the lead by some years with the creation of the Office of Navel Research. Needless to say, as the purest of all sciences, mathematical research might well have lagged behind in such an undertaking. That nothing of the sort happened is beyond any doubt traceable to one person—Mina Rees. Under her guidance, basic research in general, and especially in mathematics, received the most intelligent and wholehearted support. No greater wisdom and foresight could have been displayed and the whole postwar development of mathematical research in the United States owes an immeasurable debt to the pioneer work of the Office of Naval Research and to the alert, vigorous and farsighted policy conducted by Mina Rees. The influence of these policies has been such that it vitally affected later developments: the activities of Air Force and Ordnance research, the National Science Foundation itself. It is well known that in these more recent organizations Mina Rees was constantly appealed to for counsel and guidance.

As Mina Rees leaves her task, the Council of the American Mathematical Society desires to express to her in the name of the whole mathematical community its warmest feelings of appreciation for her past performance and extends to her its best wishes for the future.[1]

---

[1]Courtesy of the Archives of the Graduate Center of CUNY.

APPENDIX G

National Academy of Sciences Letter with regard to Public Welfare Medal and Photograph

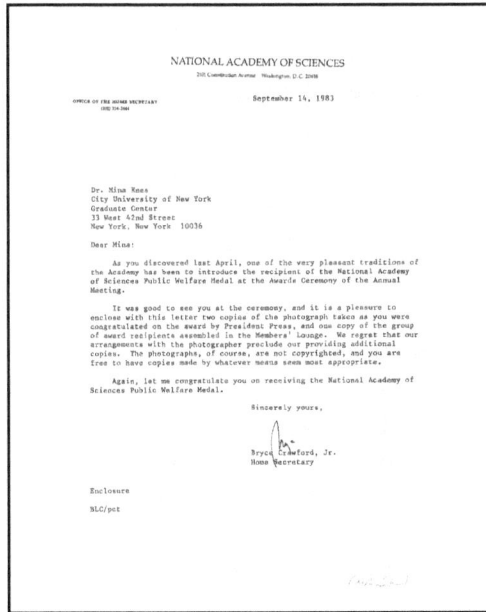

FIGURE G.1. Letter to Rees from National Academy of Sciences's Home Secretary Bryce Crawford forwarding her a photo of the 1983 Public Welfare Medals recipients. Courtesy of the Archives of the Graduate Center, CUNY.

FIGURE G.2. National Academy of Sciences 1983 Public Welfare Medal recipients. Rees front right. Courtesy of the Archives of the Graduate Center of CUNY.

APPENDIX H

Awards

King's Medal for Service in the Cause of Freedom (Britain), 1948

President's Certificate of Merit (U.S.A.), 1948

Outstanding Professional Woman of the Year, The business and Professional Women's Club of New York City, 1961

Hu and Gung Award for Distinguished Service to Mathematics, Mathematical Association of America, 1962

New York City Public Service Award for Professional Achievement, 1964

American Association of University Women Achievement Award, 1965

President's Medal, Hunter College, 1971

Alumni Medal of the University of Chicago, 1971

Elizabeth Blackwell Gold Medal of Hobart and William Smith Colleges, 1971

Chancellor's Medal, City University of New York, 1972

Public Welfare Medal. National Academy of Sciences, 1983

Distinguished Service Medal, Teachers College, Columbia University, 1983

Mayor's Award of New York City for Science and Technology, 1964 &
1986

IEEE Computer Society Pioneer Award, 1989

Honorary Degrees

Mt. Holyoke College, Sc.D., 1962

Wilson College, Sc.D., 1964

Wheaton College, Norton, MA, Sc.D., 1964

Oberlin College. Sc.D., 1964

Rutgers University, Litt.D., 1965

University of Michigan. Sc.D., 1970

Miami University, Ohio, LL.D., 1970

Columbia University, LL.D., 1971

University of Rochester, Sc.D., 1971

New York University, L.H.D., 1971

Marymount College, L.H.D., 1971

Nazareth College of Rochester, Sc.D., 1971

Carnegie Mellon University, Sc.D., 1972

University of Illinois, Sc.D., 1972

Mt. Sinai School of Medicine, Sc.D., 1972

Hunter College, Sc.D., 1973

City University of New York, 1974

Stevens Institute of Technology, Eng.D., 1980

Service

American Association for the Advancement of Science
    Vice-President and Chairman of Section A, 1954
    Board of Directors, 1957–1960, 1966–1972
        President Elect, 1970
        President, 1971
        Chairman of the Board, 1972

National Research Council, Mathematical Division, 1953–1956
    Executive Committee, 1954–1956
    Commission on Survey of Research in Mathematics in the U.S.A.
    Chairman, Sub-committee on Applied Mathematics, 1954–1956

National Science Board, 1964–1970

Advisory Panel for Mathematics, National Science Foundation, 1955–1958

Council of Graduate Schools in the United States
    Executive Committee, 1967–1971
    Chairman, 1970

New York Academy of Sciences, Council member, 1957–1960

Conference Board of the Mathematical Sciences, 1959–1962
    Vice-chairman, 1961

Mathematical Association of America
  Committee of the Undergraduate Program in Mathematics, [c. 1950]
  New York Metropolitan Section Vice-President 1955, Chairman 1956
  Vice-President 1963–1965
  Committee for The Award for Distinguished Service to Mathematics, 1959–1968
  Planning Committee for History of American Mathematics in WW II Project, 1980–1981

American Mathematical Society, Trustee, 1955–1959

Advisory Committee on Mathematics, National Bureau of Standards
  Member 1954–1958, Chairman 1954–1957

General Sciences Advisory Panel. Department of Defense, 1958–1961

Advisory Board, Computation and Exterior Ballistics Laboratory
  U.S. Naval Proving Ground, Dahlgren, 1958–1961

Commission on the Humanities, 1963–1964

Regional Committee, Woodrow Wilson Fellowship Program, 1957–1961

Honorary Advisor to the British University Summer Schools, 1957

New York State Advisory Council on Graduate Education, 1962–1972

Society for Industrial and Applied Mathematics
  Council Member, 1955–1958
  Committee on Visiting Lecturers, 1959–1960
  Representative on AAAS Council, 1958–1962
  Institute for Mathematics and Society, Board of Directors

American Conference of Academic Deans, Executive Committee, 1960–1962

School Mathematics Study Group, Advisory Panel, 1962–1965
    Summer Writing Group, 1959

Ad Hoc Committee on Education in Predominantly Negro Colleges
    American Council on Education, Chairman, 1963–1965

New York State Advisory Council on Graduate Education

Graduate Record Examinations Board

Association Hospital Service of New York and Blue Cross and Blue Shield of Greater New York, Board of Directors, 1963–1974

Health Services Improvement Fund (Blue Cross and Blue Shield of Greater New York), Board of Trustees, 1974–1983

Lenox Hill Neighborhood Association. Board of Directors, 1972–1981

Metropolitan Life Insurance Co. Educational Foundation, Advisory Committee, 1977-1982

University of Chicago, Visiting Committee member, The College, 1984-[1]

The Graduate School and University Center, CUNY, Board of Visitors, 1984-[1]

United Chapters of Phi Beta Kappa, Senator at Large
    Chairman, Committee on Visiting Scholars 1984–[1]

Woodrow Wilson National Fellowship Foundation
    Member, Board of Trustees, Executive Committee, 1984–[1]

---

[1]Termination date missing from the Rees vitae.

Association for Computing Machinery
    Founding Executive Council, 1947–48
    Nominating Committee, 1952
    Trustee, 1955-1959

Bureau of the Census, Consultant, 1960

Chronology

1902 Mina Rees born, Cleveland, Ohio
1919 Graduated valedictorian, Hunter High School, New York
1923 Graduated summa cum laude, Hunter College, New York
1925 M.A., Columbia
1931 Ph.D., University of Chicago under L. E. Dickson
1932 Professor, Hunter College
1943 Technical Aide and Executive Assistant to Warren Weaver, Applied Mathematics Panel, Office of Science Research and Development
1946 Head of Mathematics Branch, Office of Naval Research
1948 Presidential Certificate of Merit and King's Medal
1949 First Head of Mathematical Sciences Division, Office of Naval Research
1952 Deputy Science Director, Office of Naval Research
1953 Professor and Dean of Faculty, Hunter College
1955 Married Leopold Brahdy
1961 Professor and Dean of Graduate Studies, City University of New York
1962 First recipient of Mathematical Association of America Award for Distinguished Service to Mathematics
1965 American Association of University Women Achievement Award
1968 President and Provost of Graduate Division, City University of New York
1969 First President, Graduate School and University Center, City University of New York
1970 Chairman, Council of Graduate Schools in the United States
1971 First female President of the American Association for the Advancement of Science

1972 Retired as President Emeritus and Professor Emeritus, Graduate School and University Center
1977 Leopold Brahdy dies
1983 National Academy of Sciences Public Welfare Medal
1985 Mina Rees Library of Graduate School and University Center dedicated
1997 Mina Rees dies, aged 95

Bibliography

[1] A. A. Albert, *Structures of Algebras.* Providence: Amer. Math. Soc., 1964.

[2] S. A. Amitsur, "Division Algebras. A Survey," in *Contemporary Mathematics.* Providence: Amer. Math. Soc., 1982, vol. 13, pp. 3–26.

[3] W. Aspray, "Mathematics, Computers, and Other Calculating Machines," invited address, Joint Meetings of the AMS and MAA, January 18, 2000.

[4] L. Bers, "The Migration of European Mathematicians to America," in *A Century of Mathematics in America*, P. Duren, R. A. Askey, and U. C. Merzbach, Eds. Providence: Amer. Math. Soc., 1988, vol. II, pp. 231–243.

[5] J. Byers, letter to the author, April 21, 1999.

[6] ——, Mina Rees Vita.

[7] R. Courant, E. Isaacson, and M. Rees, "On the Solution of Nonlinear Hyperbolic Differential Equations by Finite Differences," *Communication on Pure and Applied Mathematics*, vol. 5, pp. 243–267, 1952.

[8] J. H. Curtiss, "A Symposium of Large Scale Digital Calculating Machinery," *Math. Tables and Other Aids to Comp.*, vol. 2, no. 18, pp. 229–238, April 1947.

[9] R. Dana and P. J. Hilton, "Mina Rees," in *Mathematical People*, D. Albers and G. L. Alexanderson, Eds. Princeton: Princeton Univ. Press, 1985, pp. 257–267.

[10] J. Dauben and A. Robinson, *The Creation of Non-standard Analysis; a Personal and Mathematical Odyssey.* Princeton: Princeton Univ. Press, 1995.

[11] L. E. Dickson, "Linear Associative Algebras and Abelian Equations," *Trans. Amer. Math. Soc.*, vol. 15, pp. 31–46, 1914.

[12] ——, *Algebras and their Arithmetics.* New York: Dover, 1923.

[13] ——, "New Division Algebra," *Trans. Amer. Math. Soc.*, vol. 28, pp. 207–234, 1926.

[14] ——, "Construction of Division Algebras," *Trans. Amer. Math. Soc.*, vol. 32, pp. 319–334, 1930.

[15] R. Donaldson, letter to the author, August 6, 1999.

[16] W. L. Duren Jr., "Graduate Student at Chicago in the Twenties," in *A Century of Mathematics in America*, P. Duren, Ed. Providence: Amer. Math. Soc., 1988, vol. II, pp. 177–182.

[17] Editor, "Association for Computing Machinery," *Math. Tables and Other Aids to Comp.*, vol. 3, no. 23, p. 216, July 1948.

[18] B. Fein, e-mail to the author, September 26, 1999.

[19] D. Fenster, "Leonard Eugene Dickson and His Work in the Theory of Algebras," Ph.D., University of Virginia, 1994.

[20] ——, "Role Modeling in Mathematics: the Case of Leonard Eugene Dickson (1874–1954)," *Historia Mathematica*, vol. 24, pp. 7–24, 1997.

[21] ——, "Leonard Eugene Dickson (1874-1954): An American Legacy in Mathematics," *The Mathematical Intelligencer*, vol. 21, no. 4, pp. 54–59, 1999.

[22] A. for Women in Mathematics, "Rees Awarded Medal," *Assoc. for Women in Math. Newsletter*, vol. 13, pp. 9–10, May–June 1983.

[23] P. Fox, "Mina Rees (1902–)," in *Women of Mathematics: A Bio-Bibliographical Sourcebook*, L. S. Grinstein and P. J. Campbell, Eds. New York: Greenwood Press, 1987, pp. 175–181.

[24] "In Memorium," Graduate School and University Center, November 16–30 1997.

[25] "Memorial," Graduate School and University Center, December 8 1997.

[26] "Remarks and Remembrances: Memorial Service," Graduate School and University Center, December 12 1997.

[27] J. Green and J. LaDuke, "Mina S. Rees: 1902–1997," *Assoc. for Women in Math. Newsletter*, vol. 28, pp. 10–12, 1998.

[28] ——, "Women in American Mathematics: A Century of Contributions," in *A Century of Mathematics in America*, P. Duren, Ed. Providence: Amer. Math. Soc., 1988, vol. II, pp. 379–398.

[29] J. Green, J. LaDuke, S. Mac Lane, and U. Merzbach, "Mina Spiegel Rees, (1902–1997)," *Notices Amer. Math. Soc.*, vol. 45, pp. 866–873, 1998.

[30] Judy and J. LaDuke, "Rees, Mina S." Amer. Math. Soc., 2008, pioneering Women in American Mathematics: The Pre-1940s PhDs. [Online]. Available: http://www.ams.org/publications/authors/books/postpub/hmath-34

[31] J. LaDuke, "The Study of Linear Associative Algebras in the United States, 1870–1927," in *Emmy Noether in Bryn Mawr*, B. Srinivasan and J. Sally, Eds. New York: Springer-Verlag, 1983, pp. 147–159.

[32] S. Mac Lane, "Mathematics at the University of Chicago: A Briel History," in *A Century of Mathematics in America*, P. Duren, Ed. Providence: Amer. Math. Soc., 1988, vol. II, pp. 127–154.

[33] ——, "The Applied Mathematics Group at Columbia in WW II," in *A Century of Mathematics in America*, P. Duren, Ed. Providence: Amer. Math. Soc., 1989, vol. III, pp. 495–516.

[34] ——, "Interview with the author, University of Chicago," January 20 1999.

[35] A. Mann, "Letter to the Editor, Math," *The Mathematical Intelligencer*, vol. 21, p. 4, November 4 1999.

[36] M. Matsushita, "A Women Mathematician and Her Contributions: Mina Spiegel Rees," Ph.D., Teachers College of Columbia University, 1998.

[37] C. McClarty, "Emmy Noether's 'Set Theoretic' Topology: From Dedekind to the rise of functors," in *The Architecture of Modern Mathematics: Essays in History and Philosophy*, J. Gray and J. Ferreirós, Eds. London: Oxford University Press, 2006, pp. 211–235.

[38] S. McMurran and J. Tattersall, "Mary Cartwright (1900–1998)," *Notices Amer. Math. Soc.*, vol. 46, pp. 214–220, February 1999.

[39] U. Merzbach, "Interview with Mina Rees," March 19 1969.

[40] S. Moite, "Mina S. Rees 1902-1996[sic] American Applied Mathematician," in *Notable Mathematicians: From Ancient Times to the Present*, R. V. Young, Ed. Gale Research, 1998, pp. 414–415, 561.

[41] I. Niven, "The Threadbare Thirties," in *A Century of Mathematics in America*, P. Duren, Ed. Providence: Amer. Math. Soc., 1988, vol. I, pp. 209–230.

[42] M. A. of America, "Award for Distinguished Service to Mathematics," *Amer. Math. Monthly*, vol. 69, pp. 185–186, February 1962.

[43] Office of Naval Research. [Online]. Available: www.onr.navy.mil

[44] L. Osen, *Women in Mathematics*. Cambridge: The MIT Press, 1974.

[45] ——, "Mina Spiegal Rees (1902–)," in *Notable Women in Mathematics*, C. Morrow and T. Perl, Eds. San Francisco: Greenwood Press, 1998, pp. 174–180.

[46] L. Owens, "Mathematicians at War: Warren Weaver and the Applied Mathematics Panel," in *History of Modern Mathematics*, D. E. Rowe and J. McCleary, Eds. New York: Academic Press, 1989, vol. III, pp. 287–305.

[47] K. C. Redmond and T. M. Smith, *Project Whirlwind: The History of a Pioneer Computer*. Bedford, MA: Digital Press, 1980.

[48] M. Rees, Mina Rees Collection, Archives of the Mina Rees Library, Graduate Center, CUNY.

[49] ——, "Division Algebras Associated with an Equation whose Group have Four Generators," *Amer. Jour. Math.*, vol. 54, no. 1, pp. 51–65, 1932.

[50] ——, "Review: *Triumph Der Mathematik* by Heinrich Dörrie," *Scripta Mathematica*, vol. 3, pp. 345–346, October 1935.

[51] ——, "Review: *The Search for Truth* by Eric Temple Bell," *Scripta Mathematica*, vol. 4, pp. 79–80, January 1936.

[52] ——, "Review: *A Semicentennial History of the American Mathematical Society. 1888–1938* by R. C. Archibald," *Scripta Mathematica*, vol. 7, pp. 121–125, January 1940.

[53] ——, "The Mathematical Program of the Office of Naval Research," *Bull. Amer. Math. Soc.*, vol. 54, pp. 1–5, January–December 1948.

[54] ——, "Professional Opprotunities in Mathematics: A Report for Students of Mathematics," *Amer. Math. Monthly*, vol. 58, no. 1, pp. 1–23, January 1951.

[55] ——, "Digital Computers – Their Nature and Use," *American Scientist*, vol. 52, pp. 328–335, April 1952.

[56] ——, "The Mathematician in Government Establishments," in *Proceedings of a Conference on Mathematics*. AMS, NRC, NSF, 1953, pp. 57–59.

[57] ——, "Mathematics and Federal Support," *Science*, vol. 119, p. 3A, May 1953.

[58] ——, "Modern Mathematics and the Gifted Student," *The Mathematics Teacher*, vol. 46, no. 6, pp. 401–406, October 1953.

[59] ——, "Digital Computers," *Amer. Math. Monthly*, vol. 62, pp. 414–423, June–July 1955.

[60] ——, "New Frontiers for Mathematics," *Pi Mu Epsilon Journal*, pp. 122–127, Fall 1955.

[61] ——, "The Impact of the Computer," *The Mathematics Teacher*, vol. 51, pp. 162–168, March 1958.

[62] ——, "Mathematics in the Marketplace," *Amer. Math. Monthly*, vol. 65, pp. 332–343, May 1958.

[63] ——, "Support for Higher Education by the Federal Government," *Amer. Math. Monthly*, vol. 68, pp. 371–378, April 1961.

[64] ——, "The Nature of Mathematics," *Science*, vol. 138, no. 3536, pp. 9–12, October 1962.

[65] ——, "The Commitment Required of a Woman Entering a Scientific Profession, (panel discussion)," in *Women and the Scientific Professions, MIT Symposium on American Science and Engineering.* Cambridge: MIT Press, 1965, pp. 43–44.

[66] ——, "The Dilemma that Faces Us," *Amer. Assoc. of Univ. Women Journal*, vol. 59, pp. 32–34, October 1965.

[67] ——, "Efforts of the Mathematical Community to Improve the Mathematics Curriculum," in *Emerging Patterns in American Higher Education*, L. Wilson, Ed. Washington, D.C.: American Council of Education, 1965, pp. 228–233.

[68] ——, *The Mathematical Sciences: A Report.* Washington, D.C.: National Academy of Sciences, 1968, published on behalf of the Committee on Support of Research in the Mathematical Sciences of the National Research Council for the Committee on Scicne and Public Policy.

[69] ——, "A Human Approach to Population Control," *Science*, vol. 173, no. 3995, p. 2, July 30 1971.

[70] ——, "Graduate Education – A Long Look," in *Graduate Education Today and Tomorrow*, L. Kent and G. Spring, Eds. Albuquerque, NM: Univ. of New Mexico Press, 1972, pp. 139–151.

[71] ——, "The Saga of American Universities: The Role of Science," *Science*, vol. 179, no. 4068, pp. 19–23, January 5 1973.

[72] ——, "The Graduate Education of Women," in *Women in Higher Education*, L. Kent and G. Spring, Eds. Washington, D.C.: American Councel on Education, 1974, pp. 178–187.

[73] ——, "The Scientist in Society: Inspiration and Obligtion," *American Scientist*, vol. 63, pp. 144–149, March–April 1975.

[74] ——, "The Ivory Tower and the Marketplace," in *On the Meaning of the University*, S. McMurrin, Ed. Salt Lake City, UT: University of Utah Press, 1976, pp. 81–101.

[75] ——, "Mathematics and the Government: Post-war Years as Augury of the Future," in *The Bicentennial Tribute to American Mathematics 1776–1976*, D. Tarwater, Ed. Washington D.C.: Math. Assoc. Amer., 1977, pp. 101–116.

[76] ——, "ONR Pioneered Government Move to Support Mathematics," *SIAM News*, vol. 11, June 1978.

[77] ——, "Biographical Letter," *Assoc. for Women in Math. Newsletter*, vol. 9, pp. 15–18, July–August 1979.

[78] ——, "The Mathematical Sciences and World War II," *Amer. Math. Monthly*, vol. 87, pp. 607–621, October 1980.

[79] ——, "The Computing Program of the Office of Naval Research, 1946–1953," *Annals of the History of Computing*, vol. 4, pp. 102–120, November 1982.

[80] ——, "The Federal Computing Machine Program," *Annals of the History of Computing*, vol. 7, pp. 156–163, April 1985.

[81] ——, "Warren Weaver," in *Biographical Memoirs.* Washington, D.C.: National Academy Press, 1987, vol. 57, pp. 492–530.

[82] ——, "The Mathematical Sciences and World War II," in *A Century of Mathematics in America*, P. Duren, Ed. Providence: Amer. Math. Soc., 1988, vol. I, pp. 275–289.

[83] C. Reid, *Neyman From Life.* New York: Springer-Verlag, 1982.

[84] ——, *Hilbert - Courant*.   New York: Springer-Verlag, 1986.

[85] M. W. Rossiter, *Women Scientists in America: Before Affirmative Action 1940-1972.* Baltimore: Johns Hopkins Press, 1995.

[86] SIAM, "Academy of Sciences Cites Mina Rees," *SIAM News*, vol. 16, pp. 1,9, July 1983.

[87] A. M. Society, "Rees Citation," *Bull. Amer. Math. Soc.*, vol. 60, pp. 134–135, March 1954.

[88] B. Stone, "Private Correspondence," Office of the Registrar, University of Chicago, 7 October 1999.

[89] H. Troop, "Interview with Mina Rees, New York," September 4 1972.

[90] ——, "Interview with Mina Rees, Smithsonian Institute, Washington, D.C." October 20 1972.

[91] F. J. Weyl, "Mina Rees, President-Elect 1970," *Science*, vol. 167, no. 3921, pp. 1149–1151, February 20 1970.

Index

www.ingramcontent.com/pod-product-compliance
Lightning Source LLC
Chambersburg PA
CBHW070807290326
41931CB00011BA/2157